Springer Series on Touch and Haptic Systems

More information about this series at https://link.springer.com/bookseries/8786

Yitian Shao

Tactile Sensing, Information, and Feedback via Wave Propagation

 Springer

Yitian Shao
Haptic Intelligence Department
Max Planck Institute for Intelligent System
Stuttgart, Baden-Württemberg, Germany

ISSN 2192-2977 ISSN 2192-2985 (electronic)
Springer Series on Touch and Haptic Systems
ISBN 978-3-030-90841-6 ISBN 978-3-030-90839-3 (eBook)
https://doi.org/10.1007/978-3-030-90839-3

This Springer imprint is published by the registered company Springer Nature Switzerland AG
The registered company address is: Gewerbestrasse 11, 6330 Cham, Switzerland

Dedicated to my grandma,
who had been waiting for me to come home.

Series Editors' Foreword

This is the nineteenth volume of *Springer Series on Touch and Haptic Systems*, which is published as a collaboration between **Springer** and the **EuroHaptics Society**.

Tactile Sensing, Information, and Feedback via Wave Propagation provides a comprehensive introduction to haptic stimuli from the point of view of mechanical wave propagation in the skin. This book is organized into eight chapters. The first part is devoted to characterizing the vibratory response of the skin at regions surrounding where the stimuli are applied. Machine learning and statistical methods have shown how these wave patterns accurately encode the modes of interaction with touched objects. The second part of this work focuses on studying spatiotemporal patterns of vibration in the hand in order to provide a hierarchical description of tactile information which is aligned with prominent anatomical features of the hand. Finally, the book focuses on a haptic display developed by using compliant liquid dielectric actuators which cover a large area of the hand. The results described are relevant to both haptic neuroscience and engineering.

This book originates from the thesis of Dr. Yitian Shao who has received the EuroHaptics award for the Best Ph.D. Thesis in 2020. This award recognizes the relevance of this work, which could open a new path to research methods and provide reliable qualitative research tools that may benefit the whole haptics community in designing ever more realistic interfaces.

Madrid, Spain   Manuel Ferre
Ulm, Germany   Marc O. Ernst
Birmingham, UK   Alan Wing
October 2021

Preface

A longstanding goal of engineering has been to realize haptic interfaces that can convey realistic sensations of touch, comparable to signals presented via visual or audio displays. Today, this ideal remains far from realization, due to the difficulty of characterizing and electronically reproducing the complex and dynamic tactile signals that are produced during even the simplest touch interactions. This book presents methods to capture whole-hand tactile signals produced during natural interactions, characterize the information content in these signals, and use the results to guide the design of new electronic devices for tactile feedback.

The first part of the book is motivated by recent findings that touch contact-elicited tactile signals can travel long distances in the hand. In order to elucidate the role of these processes in natural touch interactions, my colleagues and I first developed wearable sensing instrumentation, comprised of high channel count accelerometer arrays, for capturing whole-hand mechanical signals during manual interactions. The temporal, spatial, and frequency structures of the signals on the hand were found to vary systematically with hand interactions and could be used to identify them. These results are consistent with findings in perception research, indicating that vibrotactile signals distributed throughout the hand can transmit information regarding explored and manipulated objects.

In the second part of the book, we investigated the information content in these tactile waves using signal processing methods. Motivated by the structure in these signals, and neuroscience considerations, we sought to extract informative representations of these signals based on solutions of an optimization problem formulated as convolutional sparse coding of natural tactile signals. This yielded a dictionary of spatiotemporal primitives that provided compact descriptions of information in high-dimensional tactile signals in the whole hand, sufficient to accurately classify touched objects and interactions. The primitive patterns were organized in ways reflecting the anatomy and function of the hand and the manual activities involved.

Finally, informed by the findings about spatiotemporal distributions of touch-elicited tactile signals, we developed new methods for distributed haptic feedback adapted to the mechanics and dynamics of the skin. We developed new spatiotemporal haptic effects by using a single actuator to generate tactile stimuli with dynamically controlled spatial extent, based on the frequency-dependent damping of propagating waves in the skin. We designed a wearable haptic feedback device based on compliant liquid dielectric actuators, which can render tactile feedback with substantial displacements and forces via a large active area interfacing the skin. This yielded practical haptic technologies that can produce rich touch feedback for human–computer interaction or virtual reality.

Stuttgart, Germany   Yitian Shao
October 2021

Acknowledgements

This book is produced based on my Ph.D. thesis. I really appreciate the assistance and support from my colleagues, friends, and family members.

First, I would like to thank my advisor, Yon Visell, for guiding and supporting me through my entire academic journey. I have been working with him as a Ph.D. in the field of haptics for five years at the University of California, Santa Barbara and two years at Drexel University, and I received much valuable advice about research topics, methodologies, and communication skills. My thesis would not be completed without his instructions. I also want to thank my committee members, Katie Byl, B. S. Manjunath and Tobias Hollerer, for their encouragement and guidance in my Ph.D. qualifying exam and defense. I want to thank Katherine J. Kuchenbecker for offering me the postdoctoral researcher position and granting me the opportunity to join the Max Planck Society and the EuroHaptics Society and move my academic career forward.

Many works included in this book cannot be accomplished without the contribution from my collaborators: Vincent Hayward, Bharat Dandu, Hui Hu, James Holbery, Siyuan Ma and Sang Ho Yoon. I feel honored to have the opportunities to work with those excellent researchers and receive help from them.

Through my Ph.D. study, I learned a lot from my colleagues in the RE Touch Lab. Discussions with Gregory Reardon about technical problems, like in physics or statistics, are always constructive and inspiring. Shantonu Biswas and Thanh Nho Do influenced me with their expertise in nanoengineering and material science. Also, I am lucky to have many supportive colleagues and friends in the lab. Anzu Kawazoe can always cheer me up with her lighthearted mood, unveiling the optimistic side of life to me. Zach Wells and Harald Schafer, beside research collaborations, brought me a lot of fun outside the lab, directing me to the ocean around Santa Barbara. I also very enjoyed the company of Yufei Hao, Taku Hachisu, Simone Fani, Jasper van der Lagemaat, and Behnam Khojasteh during their visit to my lab. Starting from the time I joined the lab in Philadelphia, I have been gaining support from Marco Janko and Bin Li. I am really grateful to have all of them as my colleagues.

My friends outside the lab also helped me passing through the challenging paths toward the finish line of my Ph.D. I want to give special thanks to Mingwei Tang, my friend since childhood, who also completed his study in the US. We traveled together to many sights and cities. Those trips reduced my anxiety and rebuilt my courage to march forward. I also value the friendship with Yan Kong and Zhenyu Yang, who organized after-work entertainment in Harold Frank Hall and provided emotional support. I am grateful that my roommates, Fengqiao Sang, and Yu-Hsuan Lee, accompanied me during the year 2020. I miss and appreciate the time with Pingge Jiang, Lei Song, Tao Sun, Yaou Zhou, and Yanan Yang during my study in Philadelphia.

I appreciate the love, care, and support from my wife, Huihui Zhang, during the last two years of my Ph.D. A delightful, sweet talk with her can always relieve my stress and hearten me, motivating me to look beyond the difficulties.

Finally, I like to thank my family, especially my grandmother. I grew up bathing in her unconditional love. She was always proud of me and had the strongest belief in me, from preschool till Ph.D., granting me the faith to overcome any challenges. I deeply regret not being able to spend enough time with her at home.

This research was partially sponsored by National Science Foundation: NSF-1446752, NSF-1527709, NSF-1628831, NSF-1623459, NSF-1751348, Microsoft Research (Applied Sciences Group) and Facebook Reality Labs. The Dissertation Fellowship provided by University of California, Santa Barbara supported me in finishing the thesis.

Contents

Chapter 1
Introduction

Abstract Touch interactions are a fundamental part of our everyday life. A long-standing goal of engineering has been to realize haptic interfaces that can convey realistic sensations of touch, comparable to signals presented via visual or audio displays. Today, this ideal remains far from realization, due to the difficulty of characterizing and electronically reproducing the complex and dynamic tactile signals that are produced during even the simplest touch interactions. We ascribe the challenge to the multiplexing of submodalities in tactile perception, the wide spatiotemporal coverage and bandwidth of tactile signals, and their rich variances during active touch. As an essential step of understanding how physical tactile signals produce the sensation of touch, this book studies the signals captured during hand interactions, characterizes their information content, and utilizes the results to design new tactile feedback methods.

1.1 Motivation

Haptics refers to the science and the engineering of technologies related to the sense of touch. Haptic interactions are essential to our daily life. We touch, perceive, and interact with the environment to complete many everyday tasks. For example, object manipulation tasks require information of touch that is received by the skin [3]. Without touch sensation, even simple manual interactions, such as picking up and lighting a match, are extremely difficult or even impossible [2]. Touch sensation is also of practical importance in many critical application areas, such as in medical examinations, where it is widely used by physicians to diagnose disease via palpation.

Tactile sensation refers to the collection of touch information from the skin and associated tissues, giving rise to perception and enabling interaction. The tactile sense can be compared with other primary sensory modalities, including vision and hearing. However, when compared with video and audio displays, existing technologies for addressing the sense of touch are still in their infancy. Despite decades of research, tactile feedback technologies (electronic devices that stimulate the sense of touch), including tactile displays that might convey realistic tactile sensations (analogous to video and audio displays), are still far from realization. This is due to the

Y. Shao, *Tactile Sensing, Information, and Feedback via Wave Propagation*,
Springer Series on Touch and Haptic Systems,
https://doi.org/10.1007/978-3-030-90839-3_1

1

difficulty of characterizing and electronically reproducing tactile signals, especially the mechanical and thermal signals that facilitate tactile perception.

As one illustration of why this is the case, consider the following. Using commodity video or audio technologies, one can readily record and reproduce optical or acoustic signals, and thus convey a realistic sense that one is seeing or hearing a recorded scene. This approach cannot be applied to the sense of touch, for several reasons. First, unlike vision and audition, which involve relatively compact sensory organs, tactile stimuli are perceived by the skin, the largest and most widely distributed sensory organ of our body, with an area on the order of two square meters. It is challenging to capture tactile signals distributed over large regions of skin. Second, human tactile sensation includes many different submodalities, such as pressure/force, vibration, and temperature, and the relationship between physical signals and sensation is not fully understood. The important perceptual range of temporal variations in such signals also varies from about 1 ms to one second or longer (i.e., 0–1000 Hz in frequency). Third, the dynamical response of the skin and body across this range has a pronounced effect on the signals captured by the peripheral nervous system. Thus, it is difficult to predict how a physical stimulus will give rise to sensation. Fourth, and most importantly, touch is an active sense. Tactile signals felt by the skin are frequently extremely different event when the same object is touched slightly differently. The large effects of performed actions or active movements on touch sensation prevent tactile experiences from being captured or measured independent of the actions or movements involved.

Thus, in order to achieve the goal of engineering tactile technologies for sensing, feedback, and display that can match human abilities, it is fundamentally important to characterize physical tactile signals and their relationship with sensation. This book presents several advances motivated by these considerations. Specifically, it describes new methods for capturing dynamic, spatially distributed tactile signals in the whole hand during active touch (Chaps. 3 and 5), revealing the important role played by mechanical wave transmission in the skin. This yielded a new and surprisingly rich view of what is felt by the whole hand during touch interactions. Next, it presents an investigation of the information content in these whole-hand tactile signals, using data-driven methods informed by an efficient encoding hypothesis (Chap. 4). Further informed by these findings, it introduces new methods for providing spatiotemporal tactile feedback to the hand, which are based on the physics governing the wave phenomena uncovered in the preceding chapters (Chap. 6). Finally, it presents methods for tactile feedback using emerging soft material technologies based on the electrostatic generation of wave-like mechanical responses in a liquid medium (Chap. 7). A more detailed summary of the contents and contributions of the book are provided in the following sections.

1.2 Overview

This book is organized around several chapters presenting the context for this work and reporting the findings as reflected in the most significant published papers my collaborators and I produced.

Chapter 2 summarizes knowledge about human tactile perception, explaining how the sense of touch is produced from external mechanical and thermal stimuli applied to the skin. It also summarizes the anatomy and sensory physiology of touch, and the role of biomechanics. In addition, a brief overview of existing tactile display technologies is included in this chapter. More detailed background information is included in the central chapters of the book.

Little is known about the dynamic mechanical signals that are felt by the entire hand during active touch and how they vary with the properties of objects or the manner in which they are touched. Motivated by this, Chap. 3 (which is based on publication [4]) presents an investigation of these issues using a new sensing methodology based on a wearable accelerometer array that we created for this work. This research revealed a surprising view of touch sensing, presented for the first time in this work, associated with spatiotemporal patterns of vibration distributed in the whole hand that are elicited by naturally occurring touch contact. Analysis revealed how these patterns vary with the touch interactions and objects involved and encode attributes of touched objects and the manner in which they are touched. This research provides information that can guide the design of tactile technologies (addressed in Chaps. 5 and 6 of the book) and contributes a new view of tactile sensing and the important role played by the transmission of vibrations, as mechanical waves, in hand tissues during touch interaction.

The findings from Chap. 3 show how natural touch interactions excite complex spatiotemporal patterns of vibration in the whole hand. These tactile signals accompanying natural touch interactions are high-dimensional and vary systematically with the interactions involved, but the actual information they contain—in other words, the latent dimensionality of the high-dimensional distributed vibration signals—is unknown. This issue is investigated in Chap. 4, which is based on publications [5]. The approach is based on signal processing analyses guided by an efficient sensory information encoding hypothesis, similar to those that have been successfully applied to studies of information in visual and auditory scenes. My colleagues and we formulated the analysis as an optimization problem based on a variation on convolutional non-negative matrix factorization (a form of sparse convolutional coding). We show how optimally encoding a dataset of thousands of naturally occurring whole-hand tactile stimuli yields a compact lexicon of 5–12 primitive spatiotemporal patterns that sparsely represent information in the entire tactile dataset, enabling the tactile interactions to be classified with an accuracy exceeding 95%. To clarify the relevance of these signals to tactile sensation, we also employed a state-of-the-art model for simulating the neural responses of hundreds of tactile afferent nerve fibers in the whole hand in response to the measured signals. We show how applying an efficient encoding model to this simulated spiking neural data yields a lexicon of primitive patterns that are strikingly similar to those obtained from the efficient encoding of the mechanical data. The results suggest that the biomechanics of the hand, which is responsible for transmitting tactile signals throughout the hand, enables efficient perceptual processing by effecting a pre-neuronal (i.e., purely mechanical) compression of information in tactile signals.

The findings from Chaps. 3 and 4 were based on data collected using a new electronic sensing methodology proposed here. However, the sensor array used in those works constrained the extent to which free movements of the hand could be performed. As noted in the introduction, tactile signals vary greatly with the manual actions that are performed. To address this deficiency, Chap. 5 (based on publication [6]) presents the design of, and experiments using, an untethered, 126-channel wearable tactile sensor array for capturing the whole-hand tactile signals. The design of this sensor array was optimized in a way that can adapt to the anatomy of any adult hand, enabling unconstrained hand movements. Using custom electronics, we also designed this device so as to make it usable outside of the laboratory. The sensors in the array possess a frequency bandwidth from 0 to 700 Hz, spanning nearly the entire range of human tactile frequency sensitivity. As we demonstrate in several experiments, this system enables the real-time measurement of whole-hand tactile signals during whole-hand active touch in a regime that is not captured with existing devices, and with a comparatively large dynamic range, spatial extent, and temporal resolution.

The research presented in Chaps. 3, 4, and 5 revealed the prominent role played by tactile signals in the whole hand. These signals consist of mechanical waves in the skin that are elicited through touch contact. It has been previously observed that such wave phenomena also arise when vibrations in the skin are excited by technologies that provide vibrations to the skin. However, prior engineering research has regarded such effects as artifacts that degrade the spatial and temporal qualities of vibrations delivered by arrays of vibration transducers, such as vibrator arrays integrated in a sleeve or glove. Chapter 6 (based on publication [1]) adopts a contrary viewpoint in order to design new feedback technologies that can exploit such wave phenomena. In it, we present theoretical analyses and experiments using full-field optical vibrometry in order to identify the important effect of frequency-dependent damping on wave transmission in the hand. Informed by these observations, my collaborators and we designed novel methods for stimulating the skin with vibrations that can be provided by a single actuator and that can nonetheless excite controlled wave fields—distributed vibrations in the hand—that expand or contract in spatial extent from the location where they are applied. We present the results of perception experiments demonstrating that these applied vibrations are indeed perceived as expanding and contracting. These findings demonstrate how the physics of waves in the skin can be exploited for the design of spatiotemporal tactile effects that are practical and effective.

The next chapter of the book adopts an alternative approach for engineering tactile feedback technologies that can stimulate large areas of the skin using wave-like effects in a physical medium, rather than in the skin itself. Informed by recent findings in soft robotics research, in Chap. 7 (based on publication [7]) we present a new tactile feedback device design based on compliant, liquid dielectric actuators. The device uses a liquid medium that is compressed via the electrostatic attraction of an array of opposed hydrogel electrodes in order to hydraulically amplify mechanical displacements of a compliant pouch encapsulating the medium. This actuator is thereby able to produce substantial displacements and forces that can be achieved via a thin and compliant surface with a large active area. The intrinsic compliance of this interface also lends a comfortable quality to the feedback it provides. This technology hold promise for the engineering of new wearable haptic devices.

Chapter 8 summarizes the research in this book, synthesizes the main findings, and provides a discussion of future research directions and applications of this work.

1.3 Contributions

The key contribution of this book may be summarized as follows:

1. It presents a novel view of tactile signals in the whole hand that are produced through vibrations excited as mechanical waves during natural touch interactions (Chap. 3 and publication [4] in Proceedings of the National Academy of Sciences). This research expands our understanding of tactile function, which has hitherto been based primarily on information collected at or near to the region of contact with an object, to remote regions spreading across the whole hand, and holds implications for the design of new technologies for touch sensing and feedback. This research was widely covered in news media and online publications, and our paper has been cited in dozens of subsequent publications.
2. Related to these findings, it presents new characterizations of touch-elicited whole-hand mechanical signals and shows how the modes of interaction with touched objects are encoded in the spatiotemporal patterns of the signals (Chap. 3 and [4]). This research contributes toward a long-term scientific challenge, which is to clarify the role of distributed sensory resources in the perceptual recovery of object attributes during active touch, and provides a demonstration of the utility of remote tactile sensing for capturing touch interactions. This knowledge may inform new approaches to the engineering of robotic or prosthetic hands with integrated tactile sensing.
3. The discovery that the anatomy and biomechanics of the hand, which distribute tactile signals (as mechanical waves) far from the location of skin-object contact, produce an efficient encoding that enables tactile information to be represented in terms of a few primitive wave patterns (Chap. 4 and publication [5] in the journal Science Advances). This contribution refines our understanding of how the mechanics of hand tissues and touch contact are connected with neural processing and conscious perception. This research was widely covered in news media and online and was featured on an episode of the popular Science (AAAS) podcast in April 2020. An earlier, related paper presented at a conference was runner-up for the Best Technical Paper Award at IEEE Haptics Symposium 2016.

4. The design and development of a new method and system, comprising a 126-channel anatomically optimized wearable sensor array, can measure whole-hand tactile signals during active, free-hand interactions (Chap. 5 and publication [6] in the journal IEEE Sensors). The experiments reveal the unprecedented spatial and temporal resolution provided by this sensing approach. This method and system enable new experimental methodologies for investigating human tactile sensing and provide a model for tactile sensing that may be used with robotic and prosthetic limbs.

5. The first methods for providing haptic feedback that exploits the physics of mechanical wave transmission in the skin (Chap. 6 and publication [1] in Proc. IEEE World Haptics 2019). This method exploits the important effect of frequency-dependent damping of waves in the skin in order to enable the controlled delivery of spatiotemporal tactile patterns to the whole hand via a single vibration actuator. The method is simple, resource-efficient, and perceptually effective. Despite its simplicity, a demonstration of this method that my colleagues and I presented at IEEE World Haptics Conference 2019 was awarded by a panel of experts as the Best Technical Demonstration at the conference. Due to the hands-on nature of haptics, such demonstrations are highly valued in the haptics research community. The practicality of this method makes it especially suitable for use in emerging haptic technologies for virtual reality. In 2020, we were awarded three patents on this technology, based on a filing supported in part by an industry partner, Facebook Reality Labs.

6. A new method, the first of its kind, for stimulating distributed areas of the skin using wave-like effects provided by a compliant, hydraulically amplified electrostatic actuator array (Chap. 7 and publication [7] in Proc. IEEE Haptics Symposium 2020). The device is thin and compliant, with a distributed display area that may be scaled in size. It produces substantial displacements and forces under voltage control, making it suitable for application in emerging wearable technologies that provide distributed haptic feedback.

References

1. Dandu, B., Shao, Y., Stanley, A., Visell, Y.: Spatiotemporal haptic effects from a single actuator via spectral control of cutaneous wave propagation. In: 2019 IEEE World Haptics Conference (WHC), pp. 425–430. IEEE (2019)
2. Johansson, R.S.: The effects of anesthesia on motor skills. https://youtu.be/0LfJ3M3Kn80
3. Johansson, R.S., Häger, C., Riso, R.: Somatosensory control of precision grip during unpredictable pulling loads. Exp. Brain Res. **89**(1), 192–203 (1992)
4. Shao, Y., Hayward, V., Visell, Y.: Spatial patterns of cutaneous vibration during whole-hand haptic interactions. Proc. Natl. Acad. Sci. **113**(15), 4188–4193 (2016)
5. Shao, Y., Hayward, V., Visell, Y.: Compression of dynamic tactile information in the human hand. Sci. Adv. **6**(16), eaaz1158 (2020)

6. Shao, Y., Hu, H., Visell, Y.: A wearable tactile sensor array for large area remote vibration sensing in the hand. IEEE Sens. J. **20**(12), 6612–6623 (2020)
7. Shao, Y., Ma, S., Yoon, S.H., Visell, Y., Holbery, J.: Surfaceflow: large area haptic display via compliant liquid dielectric actuators. In: 2020 IEEE Haptics Symposium (HAPTICS), pp. 815–820. IEEE (2020)

Chapter 2
Background

Abstract This chapter summarizes general knowledge of haptic science and engineering, including an overview of haptic perception, modalities of haptic sensing, and the biomechanics of skin. In addition, it provides a brief review of existing haptic feedback solutions. This chapter also introduces background knowledge suitable for contextualizing the research presented in this book. More detailed reviews and discussions of prior literature associated with the contributions of the main chapters have been integrated with the introductory sections of Chaps. 3 through 7 that follow.

2.1 Introduction

Haptics is the field of scientific and engineering research related to the sense of touch. It encompasses research on how humans (or other animals) perceive and interact with their surroundings through touch sensing and movement, and how to engineer technologies that address the sense of touch. Scientists have studied the sense of touch for well over a century, elucidating many aspects of the biology, neuroscience, perception, and motor control related to this sense, although aspects of the sense of touch remain less understood than vision and audition. In contrast, despite decades of research, the field of haptic engineering remains less developed than engineering for the visual and audio modalities, for reasons that were explained in the introduction and further discussed in the next sections. Consequently, while a great variety of consumer technologies, including high-fidelity video and audio displays, are able to reproduce plausibly realistic experiences of seeing real or synthetic objects (through images on a computer screen, for example) or hearing sound events, few comparable haptic display devices exist, and none are yet able to even approximately match human abilities of haptic perception and action. This is partly due to the lack of understanding of the physical processes in skin tissues and neural processing involved during hand interactions and perception of touch, not only partly due to the difficulty of electronically producing mechanical signals with the many degrees of freedom and high-resolution spatial and temporal variations that are captured by the human

Y. Shao, *Tactile Sensing, Information, and Feedback via Wave Propagation*,
Springer Series on Touch and Haptic Systems,
https://doi.org/10.1007/978-3-030-90839-3_2

haptic system but also due to the integrated nature of sensing and movement that underlies haptic experiences.

The skin is our largest external sensory organ. It is innervated by several hundred thousand tactile afferent neurons that capture mechanical and thermal signals through the skin and translate them into volleys of neural spikes that are transmitted to the central nervous system [19]. As a simple example, even a light contact with an object at the fingertip elicits responses in thousands of mechanically sensitive tactile afferent neurons that innervate the skin. The complex spiking patterns of these neurons reflect the time-varying, distributed stresses or strains applied to the skin. The resulting neural spikes ascend the peripheral nervous system via the dorsal spinal column, where they form synaptic junctions with second-order afferent neurons in the brainstem (cuneate nucleus), then in the central brain region (thalamus), and subsequently reach the well-developed somatosensory (touch-related) regions of the neocortex comprising the outer layers of the brain [92]. At each of these waypoints, complex processes of neural integration combine and process the streams of tactile inputs. Recently, however, haptics researchers have noted the important role that is played by the biomechanics of the skin prior to the transduction of mechanical signals into neural information. These biomechanical effects are not fully understood, and it is this aspect of tactile sensing that is especially relevant to the area of research encompassed by this book.

The haptic sense is an active sensory modality. Most touch activities and experiences arise from the coordination of movement, or action, and sensing. Haptic perception is often decomposed into tactile or kinesthetic modalities. Tactile sensing refers to light touch sensing and is associated with touch sensations felt via skin contact. Kinesthesia refers to the sense of position and movement of the body. Kinesthetic sensations are produced by proprioceptors inside the muscles, tendons, and joints, from which we receive the information of body position and orientation, as well as an external force and torque applied to our body. Some object properties can be perceived through kinesthetic sensing, such as weight and volume of the objects, when unsupported holding and actively following the contour of the objects are involved [63]. Tactile sensations, on the other hand, are produced from the receptors inside the skin tissues. They provide information such as pressure, vibrations, and temperature stimuli applied to the skin. Properties like hardness and roughness can be obtained via skin contact with the objects when skin deformation (normal indentation and lateral stretch) and vibration signals are sensed by the mechanoreceptors widely distributed across the entire body. Tactile sensing also includes the sensation of temperature and pain, which are important but will not be discussed in this book. Instead, this book focuses mainly on tactile sensing of mechanical signals. Such signals can be thought of as distributed patterns of time-varying skin deformation (dynamic skin strain fields) elicited by physical contact with touched objects.

2.2 Human Haptics and the Hand

When humans perceive and interact with objects, we most often use our hands. Manual touch and interactions with objects produce a wealth of tactile information, through which the brain can infer many properties of the objects. In addition to haptics, other senses are often involved, but it is well established that familiar objects and their properties can be accurately identified through touch alone [59].

Tactile sensation is mediated by many thousands of mechanoreceptors—specialized end organs of the peripheral sensory nervous system. These mechanoreceptors are widely distributed in the skin and other tissues of our body. Mechanical signals captured by these mechanoreceptors produce spikes in their associated sensory neurons. The signals received by those receptors change the electrical potentials of their afferents, transmitting information to the central nervous system.

Different types of skin possess different populations of mechanoreceptors with different properties. Glabrous skin is the skin covering the palmar side of our hands and sole-facing portion of our feet. It is thus involved in most manual interactions. There are four types of mechanoreceptors with myelinated afferents innervating the glabrous skin (Table 2.1). They may be grossly distinguished by their ranges of frequency sensitivity, their receptive field sizes, and their densities [43]. Merkel cells and Meissner corpuscles are small, densely populated, shallow skin layers and possess small receptive fields. Merkel cells are sensitive to quasi-static pressure or displacement, while Meissner corpuscles are sensitive to transient contact with the skin or vibrations. Pacinian corpuscles (PCs) and Ruffini corpuscles are larger mechanoreceptors, populate deeper layers of skin and related tissues, and possess large receptive fields. PCs are exquisitely sensitive to vibrations across a wide frequency range. At frequencies of 200–300 Hz they readily respond to skin displacements on the order of 10–100 nm. These are on the order of 1000 PCs in an adult hand. Ruffini corpuscles are sensitive to lower frequency stretching of the skin, such as the mild skin stretching produced by an object lifted by the hand. Each type of mechanoreceptor is associated with a tactile afferent class: FA-I (Meissner), SA-I (Merkel), FA-II (Pacinian), or SA-II (Ruffini) (Table 2.1). They can be categorized as either fast or slow adapting types based on their response behavior to skin deformation. Fast adapting afferents are sensitive to dynamic skin deformation of relatively high frequency,

Table 2.1 Mechanoreceptors in glabrous skin of the hand [10, 12, 43, 47, 105]

Afferent type	Nerve ending	Response range (Hz)	Receptive field size (mm²)
Fast Adapting Type I (FA-I)	Meissner	~5–50	~13
Slow Adapting Type I (SA-I)	Merkel	<~5	~11
Fast Adapting Type II (FA-II)	Pacinian	~10–1000	~101
Slow Adapting Type II (SA-II)	Ruffini-like	<~5	~59

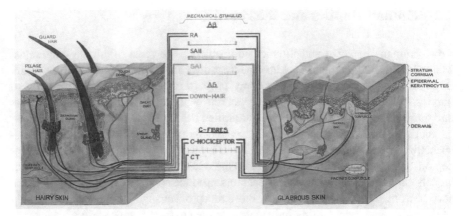

Fig. 2.1 Mechanoreceptors innervating hairy and glabrous skin. Reprinted from Neuron, 82(4), F. McGlone, J. Wessberg, and H. Olausson, Discriminative and affective touch: sensing and feeling, p. 740 [74], copyright (2014), with permission from Elsevier

while slow adapting ones are prone to respond to the lower frequency or constant skin deformation.

Like the mechanoreceptors, the afferents can also be categorized according to receptive field size, described as Type I (small) or II (large). Type I afferents terminate closer to the surface of the skin and are more concentrated toward the fingertips, as compared to Type II afferents. Type I afferents densely innervate in the fingertip, around 141 afferents per cm^2 for FA-I and 70 afferents per cm^2 for SA-I. Each afferent can innervate multiple mechanoreceptors. Their density decreases near the proximal region of the hand, around 25 and 10 afferents per cm^2, respectively. In comparison, Type II afferents are more uniformly distributed in the hand, with innervation density around 10 afferents per cm^2 [44].

Hairy skin covers most of the rest of the body (Fig. 2.1). It contains PCs, Merkel cells, Meissner corpuscles along with both hair follicle receptors and C-tactile afferents, which are not present in glabrous skin but have been associated with pleasant touch [1]. Pacinian corpuscles in hairy skin are fewer in number and located in deeper tissues [12], as compared to those in the glabrous skin. As a result, hairy skin is somewhat less vibration-sensitive (higher sensory threshold) than glabrous skin.

Despite the response behavior and innervation differences of the tactile afferents, the innervating regions of those afferents greatly overlap in the skin [44]. In the nervous system, natural stimuli elicit responses in all afferent types. The resulting neural signals are integrated at the earliest stages of neural processing. Tactile sensing of natural stimuli thus reflects an interplay of input from tactile afferents associated with different sensing modalities [92].

Perceptual acuity varies with ambient conditions and other factors. The threshold of hand vibration sensing varies under different conditions, including stimulus frequency, temperature, hand postures, and forces. Vibration frequency has a significant influence on the threshold. Studies about absolute vibration sensing threshold show a similar shape of the threshold–frequency relationship curve, across different contact conditions: flat palm on a vibrating plate, gripping a handle, and indenting fingertip [79]. The threshold curves have a U-shaped contour, with the most sensitive frequency band around 150–300 Hz [17]. Those studies also show that human hands are extremely sensitive to vibrotactile stimuli and can detect vibrations with amplitude near 0.01 μm. Acuity thresholds, measured by detection of the gap between two points on the skin, are different across the body and increase with age [101]. The hand has relatively higher spatial acuity compared to other regions of the body. Fingertips are the most sensitive region of the hand, with a threshold below 2 mm for the youths. In contrast, the palm regions are less sensitive, with a threshold above 5 mm.

2.3 Skin Biomechanics

The skin is the sensory organ of touch and possesses important biomechanical properties, including viscoelasticity and hyperelasticity. Typical mechanical parameters (elastic modulus, density, and Poisson's ratio) are reported in Table 2.2.

The skin is a distributed sensory medium whose infinitely many coupled degrees of freedom are excited in complex ways during tactile interactions. A key challenge in haptic engineering is to find methods for stimulating this continuous medium via practical devices with few mechanical degrees of freedom. Skin–object contact produces distributed and dynamic patterns of skin strain that depend on the mechanical properties of the skin. These skin strain patterns excite neural responses—volleys of spikes transmitted to the brain from thousands of mechanoreceptors. The complex geometry and continuum mechanics of the skin make it challenging to predict the neural signals that will be produced by a stimulus delivered to the skin. Biomechanical models of skin have been developed to characterize the skin behavior during touch, for example with the finger. Many groups have proposed models of finger tissue responses to mechanical stimuli [55, 84, 93, 95, 116]. The literature on skin biomechanics is large, and a full discussion would be out of scope here.

The researches in Chaps. 3, 4, 5, and 6 of this book were particularly motivated by the capacity of skin vibrations to propagate as waves away from their original location. It has long been observed that locally applied vibrations evoke mechanical waves in the skin [45, 78, 99] that travel at speeds of 4–20 m/s [71]. Processes

Table 2.2 Mechanical properties of the skin

Mechanical property	Approximate value	Reference
Elastic modulus	0.13 MPa	Khatyr et al. [54]
Density	1.02 g/cm^3	Liang and Boppart [67]
Poisson's ratio	0.5	Liang and Boppart [67]

involving low-frequency wave transmission in the skin have been modeled by several groups [57, 112, 118]. Studies suggest that mechanical waves can be transmitted through tendons [89] and bones [20] as well. The complexity and heterogeneity of skin tissues make it difficult to fully characterize these wave processes, and such models have not been previously applied in haptics.

In 1957, von Békésy reported studies of wave-like displacement patterns elicited by vibrations applied to the forearm skin, which he considered as a model system for understanding the cochlea [9] (work for which he would later receive a Nobel Prize). It has been recently observed that during naturally occurring manual activities, touching an object also excites propagating mechanical waves in the skin that reach remote locations [24, 71]. To date, the character and significance of these propagating tactile waves are not fully understood. While the transmission of vibrations in the skin has been considered to affect the performance of haptic devices [99], it is rarely accounted for in their design.

Further discussion of mechanical waves in the skin and their relevance to touch perception and interaction will be provided in Chap. 3, which presents research on whole-hand vibrations accompanying naturally occurring touch interactions, Chap. 4, which examines the information content in these signals, Chap. 5, which presents a new sensing instrument, and Chap. 6, which applies these ideas for the design of haptic feedback methods. Chapter 4 also includes a brief theoretical review of the salient wave phenomena.

2.4 Haptic Feedback Technologies

Haptic feedback technologies provide feedback to the sense of touch. Haptic interfaces integrate both sensing and feedback, reflecting the bidirectional nature of the human haptic system.

Haptic feedback plays a role in some familiar products, such as smart wristwatches, that integrate simple but elegant forms of vibration feedback for notifications, and are also present in many emerging devices or specialized instruments that are less familiar to most people. These include force feedback (kinesthetic) interfaces for simulating tissue responses in virtual surgical training, touch screen (surface haptic) interfaces that can simulate material textures via friction modulation, haptic interfaces in vehicles, and an array of emerging devices for entertainment and virtual reality. Many other research devices have been proposed during work in this field over the last half-century.

2.4.1 Kinesthetic Feedback

Kinesthetic devices are haptic interfaces that perform sensing and supply feedback, via force or movement. Impedance-type kinesthetic devices are the most common.

They sense displacement and provide force feedback. Admittance-type devices sense forces and impose displacements. They are less common, but several examples have been developed and even commercialized.

Kinesthetic devices can be characterized according to the number of actuated degrees of freedom, including one [72], two [15], three [73], six [109], or more [56]. The mechanical complexity of these devices increases rapidly with the number of degrees of freedom. Rarely have practical devices been created that actuate more than seven degrees of freedom. By comparison, the human hand has more than 20 kinematic degrees of freedom.

Several kinesthetic devices have been commercialized. They include the phantom family of stylus-based force-feedback devices [73], the force dimension omega and sigma interfaces [25], among several others. These devices are grounded (Fig. 2.2a), exerting forces between a user and a fixed surface via the interface handle, or manipulandum. Others are ungrounded (Fig. 2.2b), and exert internal forces between body segments. Several have been designed for the hand or arm; for a recent review containing many dozens of references, see [83]. The kinematic complexity of the hand makes it challenging to design hand-wearable force-feedback devices that match the kinematics of the hand. Several haptic gloves have been developed, including some that provide force feedback, but the feedback quality has typically been limited, the number of degrees of freedom typically far lower than that of the hand, and the cost quite high, due to the complexity involved, precluding their widespread adoption [85].

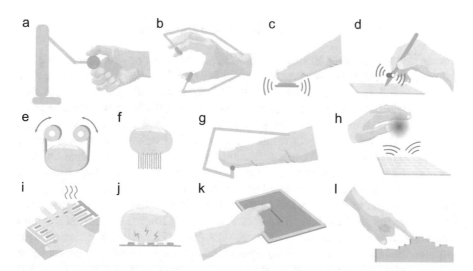

Fig. 2.2 Example setup of haptic feedback technologies. A. Grounded and B. ungrounded kinesthetic feedback. C. Vibrotactile feedback D. Tool-mediated vibrotactile feedback E. Skin stretch feedback. F. Multi-point and G. multi-DOF localized skin deformation H. Mid-air haptic feedback. I. Thermal feedback. J. Electrotactile feedback. Explorable interface based on K. friction modulation and L. shape-changing surface

Emerging research in wearable kinesthetic interfaces, including soft robotic devices, has been proposed. To date, most soft robotic devices possess extremely low control bandwidth, typically no more than 1–2 Hz. Such devices are also often imprecise, being frequently controlled in an open loop. Consequently, such devices have most frequently been applied in areas such as motor rehabilitation that require only slow actuation and simple motion patterns to be reproduced [2, 69, 87]. The natural abilities of the hand far exceed the capabilities of such devices.

2.4.2 Tactile Feedback

Tactile feedback is the feedback that is delivered directly or indirectly to the skin (e.g., through a hand-held stylus). Such feedback often takes the form of low-amplitude indentations, localized forces, or vibrations that vary in time and possibly space (in multi-actuated devices) but sometimes also through temperature or other cues. Such technologies can provide a rich variety of sensory feedback, including static and dynamic position, shape, texture, and temperature of the contact objects. Those devices can be categorized according to several feedback submodalities, including skin vibration, bulk skin stretch, localized skin deformation, lateral displacement patterns, transient contact (encounter-type displays), among others. For a recent review encompassing several dozens of references, see Culbertson et al. [21].

2.4.2.1 Vibration Feedback and Vibrotactile Technologies

Human skin is extremely sensitive to high-frequency forces or vibrations applied to the skin, often referred to as vibrotactile feedback (Fig. 2.2c). Individuals can readily perceive vibrations at frequencies near 200–300 Hz with amplitudes as low as tens of nanometers [79]. Vibrotactile feedback is very cost-effective because vibration actuators can be small and lightweight, are mechanically simple, and can be designed to be very power efficient. As a result, vibrotactile feedback techniques are widely used in consumer devices. In current consumer application, nearly all vibration actuators are designed to operate at a mechanical resonant frequency (linear resonant actuators). Older devices use technologies such as eccentric mass actuators (rotating eccentric masses driven by DC motors). The forces produced by eccentric mass actuators depend on the driving frequency. Due to these limitations, neither resonant vibration actuators nor eccentric mass actuators provide independent control over amplitude and frequency. While very power efficient, such technologies are far more limited in expressive range than other vibration actuators, such as inertial voice coil motors, that more closely resemble loudspeakers in their capabilities.

Considerable research has been undertaken on vibrotactile technologies. Such technologies include vibrotactile feedback devices as well as vibrotactile interfaces—devices integrate both sensing and feedback. The quantity of research in this area is far too large to adequately review here. A recent review describing design considerations and human factors in detail, and surveying numerous prior developments, was provided by Choi and Kuchenbecker [17].

During natural touch interactions involving a hand-held tool or probe, vibrations elicited by an interaction between a tool and an object provide ample information about the material properties and roughness of touched objects [64]. Analogous methods of vibrotactile feedback can be used to enhance the realism of virtual touch experiences, such as the hardness or roughness of virtual objects, as has been demonstrated in many studies. One interesting approach involves combining vibrotactile feedback with force feedback to simulate contact with virtual objects. A fundamental motivation for this approach arises because in haptic simulations with impedance-type force-feedback devices, bandwidth, and control stability limit the feasible stiffness of virtual objects [18]. This can cause such simulations to feel unrealistic where contacts between tools and stiff objects are involved, such as in force-feedback simulators that are used for dental training. To remedy this, several authors, including Okamura and Kuchenbecker, have investigated the use of open-loop transient vibrations to enhance such contacts [61, 81]. Such imposed vibrations can increase the apparent stiffness of virtual objects and can significantly improve the realism of object contact interactions in a virtual environment. Other researchers have investigated the use of vibrations for simulating virtual textures felt via a tool, such as a stylus on a touch screen, exploring a surface [22] (Fig. 2.2d). Vibration cues can be used to modulate frictional sensation without force feedback [60]. Recently, several researchers have also applied these methods during skin–object contact [5, 53]. Others have designed devices in which an entire surface vibrates to provide feedback that can simulate textures during bare finger-surface contact [114]. Similar methods have been used to simulate material properties via a rigid plate [106, 107], among many other methods described in the literature. The sensation of clicking a virtual button can be generated by a simple vibrotactile pulse [30], which can be adopted for making tactile interfaces for mobile devices [88]. Similar techniques have been used in consumer technologies, such as Apple's Force Touch haptic effects. Other studies have used imposed vibrations to simulate a variety of physical phenomena, such as a rolling stone inside a hand-held tube, where rolling noise and impact cues were reproduced via a vibrator inside the tube [119]. Several authors have also demonstrated how vibration feedback can be used to simulate or alter apparent objects compliance [108]. Many other applications of vibration feedback have been explored for eliciting perceptual illusions. For example, asymmetric oscillations from a vibration actuator can be designed to elicit the sensation of a tonic force pulling on the skin in one direction [4]. This effect exploits the differential sensitivity of the skin to vibrations of different frequencies. Natural haptic experiences can also be modified by processing signals felt by the skin in real time [16, 70], for modifying perceived surface properties, or potentially aiding sensory loss. By adding vibrotactile feedback to a telerobotic surgery system, tactile information of textures can be transmitted and perceived by the operator [62], which can save the manipulating force for the surgeon [75] and even improve their palpation performance [82].

Some researchers have focused on abstract or symbolic communication via vibrotactile feedback, rather than on reproducing natural tactile sensations. Some authors refer to such haptic notifications or communication units as "tactons" [6, 13]. In other applications, feedback from actuators that are located close enough to each other can be designed, using amplitude panning, onset delay sequencing, or other methods, to elicit sensations of continuous apparent motion of vibrotactile feedback moving across skin areas [3, 14, 40].

The haptic sense has often been proposed as a means of substituting for impaired visual perception, as in braille. Electronic braille display systems that use hundreds of piezo-actuated pins to translate visual information to electronically reproduced tactile patterns were developed and commercialized during the 1960s and 1970s [32]. Another emerging area of research aims to provide electronic interfaces for reproducing elements of tactile sign languages, which are widely used by deafblind individuals [26, 33]. During the last several decades, a large variety of electronic devices have also been developed for aiding visually impaired people during navigation, including many vibrotactile devices providing direction or navigation cues. Although several efforts have been made to commercialize such electronic travel aids, none have been widely adopted or used by significant numbers of people. Examples of such portable navigation devices for aiding visually impaired people include interfaces for the hand [121], arm [28], waist [36], head [42], wrist and foot [76], among other approaches [27, 35, 97]. Many vibrotactile displays have been designed for automobile drivers, including some navigation assistance devices [38].

2.4.2.2 Skin Stretch or Skin Slip

Lateral skin deformation occurs if applying tangential (shear) forces to the skin. Localized skin deformations are produced during skin–object contact during many familiar acts of object manipulation, such as lifting a drinking glass. Skin stretch is captured by numerous mechanoreceptors in the skin, and provides a dominant cue to the perceptual system about forces exerted on manipulated objects, including the weight of lifted objects. Haptic devices using skin stretch include simple devices supplying feedback at one location or at many nearby locations (as reviewed below). Many electronic devices have been developed to reproduce simple skin stretch cues at different locations of the body via an actuated mechanism, including the fingertip [31], hand [90], the forearm [11], among other locations. Skin stretch feedback can be used to produce haptic illusions of weight during virtual object manipulation [77], or to provide sensory substitution for force feedback [31, 90, 94]. Two-dimensional lateral skin deformation can also be realized via an actuated rotating ball contacting the skin [111] (Fig. 2.2e), interleaved belts [37], or a rotating disc [7], among other approaches.

2.4.2.3 Localized Skin Deformation

Several researchers have designed tactile interfaces that provide localized skin deformation via multiple actuated degrees of freedom. Over the past 50 years, many researchers have invented devices based on arrays of tens to hundreds of pins that indent nearby areas of skin [80]. However, such localized normal-indentation cues are not very perceptually effective, making these devices, which integrate many actuated degrees of freedom, inefficient. Hayward and his collaborators proposed a different approach, based on stimulating the skin via laterally imposed deformation supplied to nearby skin locations via an array of actuated elements [34, 66, 110] (Fig. 2.2f). They and others demonstrated that this technique is perceptually more effective than normal indentation. Other researchers have designed simple plates that actuate the skin at a location, such as the fingertip, in multiple directions in order to reproduce localized multi-DOF displacement cues [65] (Fig. 2.2g). Many other approaches have been described in the literature; for a recent detailed review of wearable devices adopting such approaches, see [83].

2.4.2.4 Mid-Air Tactile Feedback

Tactile sensation can be produced remotely via a mid-air haptic device, which allows free-hand interaction with a virtual object. Airborne ultrasound waves can be focused at desired locations on the skin, generating acoustic radiation force that elicits tactile sensations [39] (Fig. 2.2h). Controlling the distribution of acoustic radiation force fields can elicit the sense of volumetric shapes of virtual objects [68]. Mid-air tactile feedback has also been realized via other techniques, including air vortex [98] and air jet [102], but with lower spatial resolution compared to the feedback of ultrasound focusing.

2.4.2.5 Thermal Feedback

Thermal cues are felt during natural interactions and provide information about the material properties and temperature of touched objects. Many thermal displays have been created for various purposes using electrothermal or hydraulic systems, including Peltier or fluid-based heating and cooling systems (Fig. 2.2i). For a recent review and analysis of developments in thermal feedback technologies over the past several decades, see [46].

2.4.2.6 Electrotactile Feedback

Tactile afferent neurons innervating the skin convey the information of skin–object contact to our somatosensory cortex. Electrotactile feedback directly stimulates tactile afferent neurons using electrical current to generate tactile sensations [51] (Fig. 2.2j). Such techniques can be realized through lightweight hardware and can provide localized tactile feedback. Thus, it inspires wearable information displays for the fingertip [49, 52], the tongue [48], the forehead [50], and the arm [96].

2.4.3 Explorable Haptic Feedback

In contrast to many of the approaches described above, in which a haptic device
or interface remains in contact with the same skin area during an interaction, other
haptic devices and interfaces accommodate active exploration via hand movements,
similar to natural haptic interactions. Such devices integrate both kinesthetic behavior
(often via movement or position sensing) and tactile feedback. Many such devices
involve haptic feedback supplied via a grounded interface, rather than a wearable
instrument, and thus have a limited workspace.

A prominent category of explorable interfaces is that of surface haptic interfaces
[23]. Such devices provide feedback via an actuated touch screen or other rigid surface
explored by the hand. One prominent approach involves lateral forces produced
via electronic friction modulation. Friction modulation can be realized via friction
reduction through ultrasonic vibrations [103, 113, 115], via forces produced through
surface acoustic waves [104], or via electrostatically enhanced friction between the
skin and the surface [8, 117] (Fig. 2.2k). Further approaches are reviewed in a recent
article by Wang et al. [23].

Another approach to the design of explorable interfaces that are sometimes considered to be haptic interfaces involves shape-changing surfaces. The haptic stimuli
provided by such devices are due to the instantaneous shape of the explored surface.
Most of such devices involve many actuated degrees of freedom. Their complexity
has made them impractical for many applications. Examples include early work by
Iwata, Ishi, and their collaborators [41, 91, 120], among more recent approaches
[29, 58] (Fig. 2.2l). Recently, several authors have investigated stiffness-changing
soft interfaces that can also be explored [86, 100].

References

1. Ackerley, R., Carlsson, I., Wester, H., Olausson, H., Backlund Wasling, H.: Touch perceptions
across skin sites: differences between sensitivity, direction discrimination and pleasantness.
Front. Behav. Neurosci. **8**, 54 (2014)
2. Agarwal, P., Fox, J., Yun, Y., O'Malley, M.K., Deshpande, A.D.: An index finger exoskeleton
with series elastic actuation for rehabilitation: design, control and performance characterization. Int. J. Robot. Res. **34**(14), 1747–1772 (2015)
3. Alles, D.S.: Information transmission by phantom sensations. IEEE Trans. Man-Mach. Syst.
11(1), 85–91 (1970)
4. Amemiya, T., Gomi, H.: Distinct pseudo-attraction force sensation by a thumb-sized vibrator
that oscillates asymmetrically. In: International Conference on Human Haptic Sensing and
Touch Enabled Computer Applications, pp. 88–95. Springer (2014)
5. Ando, H., Watanabe, J., Inami, M., Sugimito, M., Maeda, T.: A fingernail-mounted tactile
display for augmented reality systems. Electron. Commun. Jpn. (Part II: Electronics) **90**(4),
56–65 (2007)
6. Azadi, M., Jones, L.A.: Evaluating vibrotactile dimensions for the design of tactons. IEEE
Trans. Haptics **7**(1), 14–23 (2014)
7. Bark, K., Wheeler, J., Shull, P., Savall, J., Cutkosky, M.: Rotational skin stretch feedback: a
wearable haptic display for motion. IEEE Trans. Haptics **3**(3), 166–176 (2010)

8. Bau, O., Poupyrev, I., Israr, A., Harrison, C.: Teslatouch: electrovibration for touch surfaces. In: Proceedings of the 23nd Annual ACM Symposium on User Interface Software and Technology, pp. 283–292 (2010)
9. v. Békésy, G.: Sensations on the skin similar to directional hearing, beats, and harmonics of the ear. J. Acoust. Soc. Amer. 29(4), 489–501 (1957)
10. Bell, J., Bolanowski, S., Holmes, M.H.: The structure and function of pacinian corpuscles: a review. Prog. Neurobiol. 42(1), 79–128 (1994)
11. Bianchi, M., Valenza, G., Serio, A., Lanata, A., Greco, A., Nardelli, M., Scilingo, E.P., Bicchi, A.: Design and preliminary affective characterization of a novel fabric-based tactile display. In: 2014 IEEE Haptics Symposium (HAPTICS), pp. 591–596. IEEE (2014)
12. Bolanowski, S.J., Gescheider, G.A., Verrillo, R.T.: Hairy skin: psychophysical channels and their physiological substrates. Somatosens. & Motor Res. 11(3), 279–290 (1994)
13. Brewster, S., Brown, L.M.: Tactons: structured tactile messages for non-visual information display. In: Proceedings of the Fifth Conference on Australasian User Interface-vol. 28, pp. 15–23. Australian Computer Society, Inc. (2004)
14. Burtt, H.E.: Tactual illusions of movement. J. Exp. Psychol. 2(5), 371 (1917)
15. Campion, G.: The pantograph mk-ii: a haptic instrument. In: The Synthesis of Three Dimensional Haptic Textures: Geometry, Control, and Psychophysics, pp. 45–58. Springer (2005)
16. Chartier, M., Thomas, N., Shao, Y., Visell, Y.: Toward a wearable tactile sensory amplification device: transfer characteristics and optimization. In: Work in Progress of the 2016 Haptic Symposium Conference, pp. 546–551. IEEE (2016)
17. Choi, S., Kuchenbecker, K.J.: Vibrotactile display: perception, technology, and applications. Proc. IEEE 101(9), 2093–2104 (2012)
18. Colgate, J.E., Grafing, P.E., Stanley, M.C., Schenkel, G.: Implementation of stiff virtual walls in force-reflecting interfaces. In: Proceedings of IEEE Virtual Reality Annual International Symposium, pp. 202 208. IEEE (1993)
19. Corniani, G., Saal, H.P.: Tactile innervation densities across the whole body. bioRxiv (2020)
20. Corso, J.F.: Bone-conduction thresholds for sonic and ultrasonic frequencies. J. Acoust. Soc. Amer. 35(11), 1738–1743 (1963)
21. Culbertson, H., Schorr, S.B., Okamura, A.M.: Haptics: the present and future of artificial touch sensation. Ann. Rev. Control Robot. Auton. Syst. 1, 385–409 (2018)
22. Culbertson, H., Unwin, J., Kuchenbecker, K.J.: Modeling and rendering realistic textures from unconstrained tool-surface interactions. IEEE Trans. Haptics 7(3), 381–393 (2014)
23. Dangxiao, W., Yuan, G., Shiyi, L., Zhang, Y., Weiliang, X., Jing, X.: Haptic display for virtual reality: progress and challenges. Virtual Real. & Intell. 1(2), 136–162 (2019)
24. Delhaye, B., Hayward, V., Lefèvre, P., Thonnard, J.L.: Texture-induced vibrations in the forearm during tactile exploration. Front. Behav. Neurosci. 6, 37 (2012)
25. Dimension, F.: Omega-7 overview. http://www.forcedimension.com/omega7-overview (2013)
26. Duvernoy, B., Topp, S., Hayward, V.: "hapticomm", a haptic communicator device for deafblind communication. In: International AsiaHaptics Conference, pp. 112–115. Springer (2018)
27. Elliott, L.R., van Erp, J., Redden, E.S., Duistermaat, M.: Field-based validation of a tactile navigation device. IEEE Trans. Haptics 3(2), 78–87 (2010)
28. Filgueiras, T.S., Lima, A.C.O., Baima, R.L., Oka, G.T.R., Cordovil, L.A.Q., Bastos, M.P.: Vibrotactile sensory substitution on personal navigation: Remotely controlled vibrotactile feedback wearable system to aid visually impaired. In: 2016 IEEE International Symposium on Medical Measurements and Applications (MeMeA), pp. 1–5. IEEE (2016)
29. Follmer, S., Leithinger, D., Olwal, A., Hogge, A., Ishii, H.: Inform: dynamic physical affordances and constraints through shape and object actuation. In: Uist, vol. 13, pp. 2501988–2502032 (2013)
30. Fukumoto, M., Sugimura, T.: Active click: tactile feedback for touch panels. In: CHI'01 extended abstracts on Human factors in computing systems, pp. 121–122 (2001)

31. Gleeson, B.T., Horschel, S.K., Provancher, W.R.: Design of a fingertip-mounted tactile display with tangential skin displacement feedback. IEEE Trans. Haptics 3(4), 297–301 (2010)
32. Goldish, L.H., Taylor, H.E.: The optacon: a valuable device for blind persons. J. Vis. Impair. & Blind. 68(2), 49–56 (1974)
33. Gollner, U., Bieling, T., Joost, G.: Mobile lorm glove: introducing a communication device for deaf-blind people. In: Proceedings of the Sixth International Conference on Tangible, Embedded and Embodied Interaction, pp. 127–130 (2012)
34. Hayward, V., Cruz-Hernandez, M.: Tactile display device using distributed lateral skin stretch. In: Proceedings of the Haptic Interfaces for Virtual Environment and Teleoperator Systems Symposium, vol. 69, pp. 1309–1314. Citeseer (2000)
35. Hesch, J.A., Roumeliotis, S.I.: Design and analysis of a portable indoor localization aid for the visually impaired. Int. J. Robot. Res. 29(11), 1400–1415 (2010)
36. Heuten, W., Henze, N., Boll, S., Pielot, M.: Tactile wayfinder: a non-visual support system for wayfinding. In: Proceedings of the 5th Nordic Conference on Human-Computer Interaction: Building Bridges, pp. 172–181 (2008)
37. Ho, C., Kim, J., Patil, S., Goldberg, K.: The slip-pad: a haptic display using interleaved belts to simulate lateral and rotational slip. In: 2015 IEEE World Haptics Conference (WHC), pp. 189–195. IEEE (2015)
38. Hogema, J.H., De Vries, S.C., Van Erp, J.B., Kiefer, R.J.: A tactile seat for direction coding in car driving: field evaluation. IEEE Trans. Haptics 2(4), 181–188 (2009)
39. Hoshi, T., Takahashi, M., Iwamoto, T., Shinoda, H.: Noncontact tactile display based on radiation pressure of airborne ultrasound. IEEE Trans. Haptics 3(3), 155–165 (2010)
40. Israr, A., Poupyrev, I.: Tactile brush: drawing on skin with a tactile grid display. In: Proceedings of the SIGCHI Conference on Human Factors in Computing Systems, pp. 2019–2028 (2011)
41. Iwata, H., Yano, H., Nakaizumi, F., Kawamura, R.: Project feelex: adding haptic surface to graphics. In: Proceedings of the 28th Annual Conference on Computer Graphics and Interactive Techniques, pp. 469–476 (2001)
42. de Jesus Oliveira, V.A., Brayda, L., Nedel, L., Maciel, A.: Designing a vibrotactile head-mounted display for spatial awareness in 3d spaces. IEEE Trans. Visual Comput. Graphics 23(4), 1409–1417 (2017)
43. Johansson, R.S., Flanagan, J.R.: Coding and use of tactile signals from the fingertips in object manipulation tasks. Nat. Rev. Neurosci. 10(5), 345–359 (2009)
44. Johansson, R.S., Vallbo, A.: Tactile sensibility in the human hand: relative and absolute densities of four types of mechanoreceptive units in glabrous skin. J. Physiol. 286(1), 283–300 (1979)
45. Johansson, R.S., Vallbo, Å.B.: Tactile sensory coding in the glabrous skin of the human hand. Trends Neurosci. 6, 27–32 (1983)
46. Jones, L.A.: Perspectives on the evolution of tactile, haptic, and thermal displays. Presence: Teleoperators Virtual Env. 25(3), 247–252 (2016)
47. Jones, L.A., Sarter, N.B.: Tactile displays: guidance for their design and application. Hum. Factors 50(1), 90–111 (2008)
48. Kaczmarek, K.A.: The tongue display unit (tdu) for electrotactile spatiotemporal pattern presentation. Sci. Iranica 18(6), 1476–1485 (2011)
49. Kaczmarek, K.A., Haase, S.J.: Pattern identification and perceived stimulus quality as a function of stimulation waveform on a fingertip-scanned electrotactile display. IEEE Trans. Neural Syst. Rehabil. Eng. 11(1), 9–16 (2003)
50. Kajimoto, H., Kanno, Y., Tachi, S.: Forehead electro-tactile display for vision substitution. In: Proceedings of the EuroHaptics (2006)
51. Kajimoto, H., Kawakami, N., Maeda, T., Tachi, S.: Tactile feeling display using functional electrical stimulation. In: Proceedings of the 1999 ICAT, p. 133 (1999)
52. Kajimoto, H., Kawakami, N., Tachi, S., Inami, M.: Smarttouch: electric skin to touch the untouchable. IEEE Comput. Graphics Appl. 24(1), 36–43 (2004)
53. Kawazoe, A., Di Luca, M., Visell, Y.: Tactile echoes: a wearable system for tactile augmentation of objects. In: 2019 IEEE World Haptics Conference (WHC), pp. 359–364. IEEE (2019)

54. Khatyr, F., Imberdis, C., Vescovo, P., Varchon, D., Lagarde, J.M.: Model of the viscoelastic behaviour of skin in vivo and study of anisotropy. Skin Res. Technol. **10**(2), 96–103 (2004)
55. Khojasteh, B., Janko, M., Visell, Y.: Complexity, rate, and scale in sliding friction dynamics between a finger and textured surface. Sci. Rep. **8**(1), 1–10 (2018)
56. Kim, S., Hasegawa, S., Koike, Y., Sato, M.: Tension based 7-dof force feedback device: Spidar-g. In: Proceedings IEEE Virtual Reality 2002, pp. 283–284. IEEE (2002)
57. Kirkpatrick, S.J., Duncan, D.D., Fang, L.: Low-frequency surface wave propagation and the viscoelastic behavior of porcine skin. J. Biomed. Opt. **9**(6), 1311–1320 (2004)
58. Klare, S., Peer, A.: The formable object: a 24-degree-of-freedom shape-rendering interface. IEEE/ASME Trans. Mechatron. **20**(3), 1360–1371 (2014)
59. Klatzky, R.L., Lederman, S.J., Metzger, V.A.: Identifying objects by touch: An "expert system." Percept. & Psychophys. **37**(4), 299–302 (1985)
60. Konyo, M., Yamada, H., Okamoto, S., Tadokoro, S.: Alternative display of friction represented by tactile stimulation without tangential force. In: International Conference on Human Haptic Sensing and Touch Enabled Computer Applications, pp. 619–629. Springer (2008)
61. Kuchenbecker, K.J., Fiene, J., Niemeyer, G.: Improving contact realism through event-based haptic feedback. IEEE Trans. Visual Comput. Graphics **12**(2), 219–230 (2006)
62. Kuchenbecker, K.J., Gewirtz, J., McMahan, W., Standish, D., Martin, P., Bohren, J., Mendoza, P.J., Lee, D.I.: Verrotouch: high-frequency acceleration feedback for telerobotic surgery. In: International Conference on Human Haptic Sensing and Touch Enabled Computer Applications, pp. 189–196. Springer (2010)
63. Lederman, S.J., Klatzky, R.L.: Hand movements: a window into haptic object recognition. Cogn. Psychol. **19**(3), 342–368 (1987)
64. Lederman, S.J., Klatzky, R.L., Hamilton, C.L., Ramsay, G.I.: Perceiving surface roughness via a rigid probe: effects of exploration speed and mode of touch (1999)
65. Leonardis, D., Solazzi, M., Bortone, I., Frisoli, A.: A wearable fingertip haptic device with 3 dof asymmetric 3-rsr kinematics. In: 2015 IEEE World Haptics Conference (WHC), pp. 388–393. IEEE (2015)
66. Lévesque, V., Pasquero, J., Hayward, V.: Braille display by lateral skin deformation with the stress2 tactile transducer. In: Second Joint EuroHaptics Conference and Symposium on Haptic Interfaces for Virtual Environment and Teleoperator Systems (WHC'07), pp. 115–120. IEEE (2007)
67. Liang, X., Boppart, S.A.: Biomechanical properties of in vivo human skin from dynamic optical coherence elastography. IEEE Trans. Biomed. Eng. **57**(4), 953–959 (2010)
68. Long, B., Seah, S.A., Carter, T., Subramanian, S.: Rendering volumetric haptic shapes in mid-air using ultrasound. ACM Trans. Graphics (TOG) **33**(6), 1–10 (2014)
69. Ma, Z., Ben-Tzvi, P.: Design and optimization of a five-finger haptic glove mechanism. J. Mech. Robot. **7**(4) (2015)
70. Maeda, T., Peiris, R., Nakatani, M., Tanaka, Y., Minamizawa, K.: Wearable haptic augmentation system using skin vibration sensor. In: Proceedings of the 2016 Virtual Reality International Conference, pp. 1–4 (2016)
71. Manfredi, L.R., Baker, A.T., Elias, D.O., Dammann III, J.F., Zielinski, M.C., Polashock, V.S., Bensmaia, S.J.: The effect of surface wave propagation on neural responses to vibration in primate glabrous skin. PloS one **7**(2) (2012)
72. Martinez, M.O., Morimoto, T.K., Taylor, A.T., Barron, A.C., Pultorak, J.A., Wang, J., Calasanz-Kaiser, A., Davis, R.L., Blikstein, P., Okamura, A.M.: 3-d printed haptic devices for educational applications. In: 2016 IEEE Haptics Symposium (HAPTICS), pp. 126–133. IEEE (2016)
73. Massie, T.H., Salisbury, J.K., et al.: The phantom haptic interface: a device for probing virtual objects. In: Proceedings of the ASME Winter Annual Meeting, Symposium on Haptic Interfaces for Virtual Environment and Teleoperator Systems, vol. 55, pp. 295–300. Chicago, IL (1994)
74. McGlone, F., Wessberg, J., Olausson, H.: Discriminative and affective touch: sensing and feeling. Neuron **82**(4), 737–755 (2014)

75. McMahan, W., Bark, K., Gewirtz, J., Standish, D., Martin, P.D., Kunkel, J.A., Lilavois, M., Wedmid, A., Lee, D.I., Kuchenbecker, K.J.: Tool vibration feedback may help expert robotic surgeons apply less force during manipulation tasks. In: Yang, G.Z., Darzi, A. (eds.) Proceedings of the Hamlyn Symposium on Medical Robotics, p. 3738 (2011)

76. Meier, A., Matthies, D.J., Urban, B., Wettach, R.: Exploring vibrotactile feedback on the body and foot for the purpose of pedestrian navigation. In: Proceedings of the 2nd international Workshop on Sensor-based Activity Recognition and Interaction, pp. 1–11 (2015)

77. Minamizawa, K., Fukamachi, S., Kajimoto, H., Kawakami, N., Tachi, S.: Gravity grabber: wearable haptic display to present virtual mass sensation. In: ACM SIGGRAPH 2007 Emerging Technologies, pp. 8–es (2007)

78. Moore, T.J.: A survey of the mechanical characteristics of skin and tissue in response to vibratory stimulation. IEEE Trans. Man-Mach. Syst. **11**(1), 79–84 (1970)

79. Morioka, M., Griffin, M.J.: Thresholds for the perception of hand-transmitted vibration: dependence on contact area and contact location. Somatosens. & Motor Res. **22**(4), 281–297 (2005)

80. Moy, G., Wagner, C., Fearing, R.S.: A compliant tactile display for teletaction. In: Proceedings 2000 ICRA. Millennium Conference. IEEE International Conference on Robotics and Automation. Symposia Proceedings (Cat. No. 00CH37065), vol. 4, pp. 3409–3415. IEEE (2000)

81. Okamura, A.M., Cutkosky, M.R., Dennerlein, J.T.: Reality-based models for vibration feedback in virtual environments. IEEE/ASME Trans. Mechatron. **6**(3), 245–252 (2001)

82. Pacchierotti, C., Prattichizzo, D., Kuchenbecker, K.J.: Cutaneous feedback of fingertip deformation and vibration for palpation in robotic surgery. IEEE Trans. Biomed. Eng. **63**(2), 278–287 (2015)

83. Pacchierotti, C., Sinclair, S., Solazzi, M., Frisoli, A., Hayward, V., Prattichizzo, D.: Wearable haptic systems for the fingertip and the hand: taxonomy, review, and perspectives. IEEE Trans. Haptics **10**(4), 580–600 (2017)

84. Pawluk, D.T., Howe, R.D.: Dynamic lumped element response of the human fingerpad (1999)

85. Perret, J., Vander Poorten, E.: Touching virtual reality: a review of haptic gloves. In: ACTUATOR 2018; 16th International Conference on New Actuators, pp. 1–5. VDE (2018)

86. Phung, H., Hoang, P.T., Nguyen, C.T., Nguyen, T.D., Jung, H., Kim, U., Choi, H.R.: Interactive haptic display based on soft actuator and soft sensor. In: 2017 IEEE/RSJ International Conference on Intelligent Robots and Systems (IROS), pp. 886–891. IEEE (2017)

87. Polygerinos, P., Wang, Z., Galloway, K.C., Wood, R.J., Walsh, C.J.: Soft robotic glove for combined assistance and at-home rehabilitation. Robot. Auton. Syst. **73**, 135–143 (2015)

88. Poupyrev, I., Maruyama, S.: Tactile interfaces for small touch screens. In: Proceedings of the 16th Annual ACM Symposium on User Interface Software and Technology, pp. 217–220 (2003)

89. Pourcelot, P., Defontaine, M., Ravary, B., Lemâtre, M., Crevier-Denoix, N.: A non-invasive method of tendon force measurement. J. Biomech. **38**(10), 2124–2129 (2005)

90. Provancher, W.R., Sylvester, N.D.: Fingerpad skin stretch increases the perception of virtual friction. IEEE Trans. Haptics **2**(4), 212–223 (2009)

91. Rosen, D., Nguyen, A., Wang, H.: On the geometry of low degree-of-freedom digital clay human-computer interface devices. In: ASME 2003 International Design Engineering Technical Conferences and Computers and Information in Engineering Conference, pp. 1135–1144. American Society of Mechanical Engineers Digital Collection (2003)

92. Saal, H.P., Bensmaia, S.J.: Touch is a team effort: interplay of submodalities in cutaneous sensibility. Trends Neurosci. **37**(12), 689–697 (2014)

93. Sakurai, T., Shinoda, H., Konyo, M.: Sharp tactile sensation using superposition of vibrotactile stimuli in different phases. In: 2013 World Haptics Conference (WHC), pp. 235–240. IEEE (2013)

94. Schorr, S.B., Quek, Z.F., Romano, R.Y., Nisky, I., Provancher, W.R., Okamura, A.M.: Sensory substitution via cutaneous skin stretch feedback. In: 2013 IEEE International Conference on Robotics and Automation, pp. 2341–2346. IEEE (2013)

95. Serina, E.R., Mote, C., Jr., Rempel, D.: Force response of the fingertip pulp to repeated compression-effects of loading rate, loading angle and anthropometry. J. Biomech. **30**(10), 1035–1040 (1997)

96. Shin, H., Watkins, Z., Huang, H.H., Zhu, Y., Hu, X.: Evoked haptic sensations in the hand via non-invasive proximal nerve stimulation. J. Neural Eng. 15(4), 046005 (2018)

97. Siu, A.F., Sinclair, M., Kovacs, R., Ofek, E., Holz, C., Cutrell, E.: Virtual reality without vision: a haptic and auditory white cane to navigate complex virtual worlds. In: Proceedings of the 2020 CHI Conference on Human Factors in Computing Systems, pp. 1–13 (2020)

98. Sodhi, R., Poupyrev, I., Glisson, M., Israr, A.: Aireal: interactive tactile experiences in free air. ACM Trans. Graphics (TOG) **32**(4), 1–10 (2013)

99. Sofia, K.O., Jones, L.: Mechanical and psychophysical studies of surface wave propagation during vibrotactile stimulation. IEEE Trans. Haptics **6**(3), 320–329 (2013)

100. Stanley, A.A., Okamura, A.M.: Controllable surface haptics via particle jamming and pneumatics. IEEE Trans. Haptics **8**(1), 20–30 (2015)

101. Stevens, J.C., Choo, K.K.: Spatial acuity of the body surface over the life span. Somatosens. & Motor Res. **13**(2), 153–166 (1996)

102. Suzuki, Y., Kobayashi, M.: Air jet driven force feedback in virtual reality. IEEE Comput. Graphics Appl. **25**(1), 44–47 (2005)

103. Takasaki, M., Kotani, H., Mizuno, T., Nara, T.: Transparent surface acoustic wave tactile display. In: 2005 IEEE/RSJ International Conference on Intelligent Robots and Systems, pp. 3354–3359. IEEE (2005)

104. Takasaki, M., Tamon, R., Kotani, H., Mizuno, T.: Pen tablet type surface acoustic wave tactile display integrated with visual information. In: 2008 IEEE International Conference on Mechatronics and Automation, pp. 357–362. IEEE (2008)

105. Vallbo, A.B., Johansson, R.S., et al.: Properties of cutaneous mechanoreceptors in the human hand related to touch sensation. Hum. Neurobiol. 3(1), 3–14 (1984)

106. Visell, Y., Cooperstock, J.R.: Design of a vibrotactile display via a rigid surface. In: 2010 IEEE Haptics Symposium, pp. 133–140. IEEE (2010)

107. Visell, Y., Cooperstock, J.R., Giordano, B.L., Franinovic, K., Law, A., McAdams, S., Jathal, K., Fontana, F.: A vibrotactile device for display of virtual ground materials in walking. In: International Conference on Human Haptic Sensing and Touch Enabled Computer Applications, pp. 420–426. Springer (2008)

108. Visell, Y., Okamoto, S.: Vibrotactile sensation and softness perception. In: Multisensory Softness, pp. 31–47. Springer (2014)

109. Vu, M.H., Na, U.J.: A new 6-dof haptic device for teleoperation of 6-dof serial robots. IEEE Trans. Instrum. Meas. **60**(11), 3510–3523 (2011)

110. Wang, Q., Hayward, V.: Biomechanically optimized distributed tactile transducer based on lateral skin deformation. Int. J. Robot. Res. **29**(4), 323–335 (2010)

111. Webster, R.J., III., Murphy, T.E., Verner, L.N., Okamura, A.M.: A novel two-dimensional tactile slip display: design, kinematics and perceptual experiments. ACM Trans. Appl. Percept. (TAP) **2**(2), 150–165 (2005)

112. Welcome, D.E., Dong, R.G., Xu, X.S., Warren, C., McDowell, T.W., Wu, J.Z.: An examination of the vibration transmissibility of the hand-arm system in three orthogonal directions. Int. J. Ind. Ergon. **45**, 21–34 (2015)

113. Wiertlewski, M., Leonardis, D., Meyer, D.J., Peshkin, M.A., Colgate, J.E.: A high-fidelity surface-haptic device for texture rendering on bare finger. In: International Conference on Human Haptic Sensing and Touch Enabled Computer Applications, pp. 241–248. Springer (2014)

114. Wiertlewski, M., Lozada, J., Hayward, V.: The spatial spectrum of tangential skin displacement can encode tactual texture. IEEE Trans. Rob. **27**(3), 461–472 (2011)

115. Winfield, L., Glassmire, J., Colgate, J.E., Peshkin, M.: T-pad: Tactile pattern display through variable friction reduction. In: Second Joint EuroHaptics Conference and Symposium on Haptic Interfaces for Virtual Environment and Teleoperator Systems (WHC'07), pp. 421–426. IEEE (2007)

116. Wu, J.Z., Krajnak, K., Welcome, D.E., Dong, R.G.: Three-dimensional finite element simu-
     lations of the dynamic response of a fingertip to vibration. J. Biomech. Eng. **130**(5) (2008)
117. Xu, H., Peshkin, M.A., Colgate, J.E.: Ultrashiver: lateral force feedback on a bare fingertip
     via ultrasonic oscillation and electroadhesion. IEEE Trans. Haptics **12**(4), 497–507 (2019)
118. Xu, X.S., Welcome, D.E., McDowell, T.W., Wu, J.Z., Wimer, B., Warren, C., Dong, R.G.:
     The vibration transmissibility and driving-point biodynamic response of the hand exposed to
     vibration normal to the palm. Int. J. Ind. Ergon. **41**(5), 418–427 (2011)
119. Yao, H.y., Hayward, V.: An experiment on length perception with a virtual rolling stone. In:
     Proceedings of Eurohaptics, pp. 325–330 (2006)
120. Yao, L., Niiyama, R., Ou, J., Follmer, S., Della Silva, C., Ishii, H.: Pneui: pneumatically
     actuated soft composite materials for shape changing interfaces. In: Proceedings of the 26th
     Annual ACM Symposium on User Interface Software and Technology, pp. 13–22 (2013)
121. Yatani, K., Banovic, N., Truong, K.: Spacesense: representing geographical information to
     visually impaired people using spatial tactile feedback. In: Proceedings of the SIGCHI Con-
     ference on Human Factors in Computing Systems, pp. 415–424 (2012)

Chapter 3
Spatial Patterns of Whole-Hand Cutaneous Vibration During Active Touch

Abstract In order to engineer haptic technologies that might provide realistic touch sensations, knowledge about what signals are felt by the hand during natural interactions is needed. Prior research has highlighted the importance of vibrotactile signals during haptic interactions, but little is known of how vibrations propagate throughout the hand. Furthermore, the extent to which the patterns of vibrations reflect the nature of the objects that are touched, and how they are touched, is unknown. We investigated the propagation patterns of cutaneous vibration in the hand during interactions with touched objects. Using an apparatus comprised of an array of accelerometers, we mapped and analyzed spatial distributions of vibrations propagating in the skin of the dorsal region of the hand during active touch, grasping, and manipulation tasks. We found these spatial patterns of vibration to vary systematically with touch interactions and determined that it is possible to use these data to decode the modes of interaction with touched objects. The observed vibration patterns evolved rapidly in time, peaking in intensity within a few milliseconds, fading within 20–30 ms, and yielding interaction-dependent distributions of energy in frequency bands that span the range of vibrotactile sensitivity. These results are consistent with findings in perception research that indicate that vibrotactile information distributed throughout the hand can transmit information regarding explored and manipulated objects. The results may further clarify the role of distributed sensory resources in the perceptual recovery of object attributes during active touch, may guide the development of approaches to robotic sensing, and could have implications for the rehabilitation of the upper extremity.

Disclaimer

Previously published as: Y. Shao, V. Hayward, and Y. Visell, Spatial Patterns of Cutaneous Vibration During Whole-Hand Haptic Interactions. *Proceedings of the National Academy of Sciences*, Apr 2016, 113 (15), 4188–4193; DOI: 10.1073/pnas. 1520866113. Reproduced here by permission of NAS.

3.1 Introduction

When we touch an object, a cascade of mechanical events ensues, and through it, vibration is transmitted, not just to the fingertips, but broadly within the hard and soft tissues of the hand. Prior research has shed light on mechanical stimuli generated during object palpation or manipulation, the transduction of such signals into neural signals, and the salience of different contact-generated stimuli. It has been shown that the responses of somatosensory neurons should be understood in light of perceptual functions that integrate input from several tactile submodalities [16, 28].

Tactile mechanics yield numerous perceptual cues that inform the brain about key properties of the external mechanical world such as the presence of an object through contact [13], slip against a surface [8], object deformation [2, 34], and object shape [25, 36]. Among these cues, touch-induced vibrations play important roles. Until recently, it has been assumed that perceptual information generated during haptic interaction is confined to the region of skin–object contact. It has been subsequently demonstrated, however, that perceptually meaningful mechanical energy can propagate away from the origin of contact, sometimes beyond the hand itself [7, 33], and that humans are capable of utilizing this information to evaluate surface roughness [18]. Recent measurements have demonstrated that skin vibrations reflect the fine-scale topography of touched objects [22]. It is nonetheless not known whether touch-elicited vibrations contain more general information about an object that would be available at significant distances from the contact location.

At frequencies greater than 100 Hz, mechanical damping dominates elasticity [35], and the skin can be thought of as a fluid-filled layer that can be excited vibromechanically [14]. In this regime, mechanical transients propagate within glabrous skin at fast, yet frequency-dependent, speeds ranging from 5 to 7 m/s within the vibrotactile range [21]. Despite the dispersive nature of wave propagation in the skin, complex waveforms appear to be well preserved at distances of at least several centimeters, and possibly much further [7], suggesting that perceptual information content may remain intact far from the site of stimulation. Although the amplitude of vibrations propagating in skin decay with distance [21, 31], decay is lower at frequencies relevant to vibrotactile sensation (250 Hz), and contact-induced vibrations can remain above detectable thresholds at distances spanning most of the hand. However, the spatial and temporal propagation patterns of touch-elicited vibrations in the hand have not been characterized.

Prior literature sheds little light on the functional role that is played by mechanoreceptors that are far removed from areas of skin that are in contact with objects, but mechanical stimuli are known to excite sensory cells over wide areas [13]. Pacinian corpuscles (PCs) have receptive fields that can span several centimeters and are located in the deep dermis of the volar (glabrous) and dorsal (hairy) skin of the hand [5, 9, 12, 15, 17, 32]. PC units respond to stimuli in a wide frequency range ($\sim$20–1 kHz). Several studies have associated PC units in hairy skin with vibrotactile sensory function [19, 23], including the detection of remote tapping [9]. That the PC system is strongly implicated in the detection of fast mechanical signals does not exclude that other populations of sensory cells may also contribute. Merkel cellneurites, which are abundant at the interface of the epidermis and are described as forming the slowing adapting receptor population, have been shown to respond to frequencies in the whole tactile frequency range [11, 20]. Similarly, the numerous Meissner corpuscles found in the epidermal grooves of the glabrous skin, and associated with fast adapting afferent units, cannot be excluded from responding to the low-frequency range (10–200 Hz) of stimuli propagating in the skin [10]. These stimuli may also provide an input to the network of tendons in the hand and associated muscle spindles [4, 24, 27].

Vibromechanical stimuli have occasionally been used in psychophysical studies on stimulus localization [6, 26]. The distribution and properties of sensory cells can also enable the remote detection of propagating vibrations away from the site of contact [7]. These processes have thus far received limited attention, but further insight into mechanisms of remote tactile sensing could shed light on sensory specializations in the whole hand. Toward this end, we developed an apparatus consisting of an array of accelerometers capable of capturing cutaneous vibration at length and time scales matched to the receptive field sizes and frequency selectivity of fast adapting cutaneous mechanoreceptive afferents in the dorsal surface of the hand. Because this device is worn on the hand, it allowed us to collect data as subjects actively touched objects. We used it to accurately map vibration propagation in the hand during active touch, grasping, and manipulation tasks.

3.2 Capturing Whole-Hand Tactile Signals

3.2.1 Apparatus

The apparatus was a customized array of 15 or 30 three-channel miniature accelerometers (model ADXL335; Analog Devices) attached to the skin (Fig. 3.1). These devices had low, but nonzero, mass (40.0 mg), wide frequency bandwidth (0–1600 Hz in X and Y; 0–550 Hz in Z), high dynamic range (−35.3 to 35.3 m/s^2), and were soldered to miniature two-sided printed circuit boards (width 6 mm, length 8 mm). They were connected to the data acquisition board via lightweight ribbon cables, whose softness helped to minimize artifacts. Ribbon cables were used to connect the accelerometers to the data acquisition board due to their softness that prevents the vibration interference among cables and accelerometers. The contact surface between the skin and the accelerometer was 4 × 4 mm in size. The analog signals were digitized with 12-bit resolution using custom electronics and were sampled at a frequency of 2.0 kHz by a data acquisition board (model PCIE-6321; National Instruments). The accelerometers were attached to the skin using a prosthetic adhesive (Pros-Aide; FXWarehouse) that ensured a consistent flexible bond over a small contact patch. The sensors were placed to provide coverage of all five fingers and dorsal surface of the hand, in correspondence with the distal, intermediate, and prox-

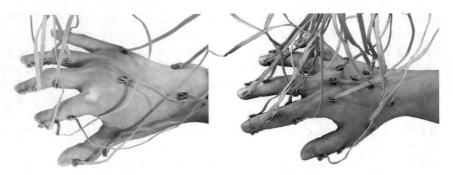

Fig. 3.1 15 (left) and 30 (right) accelerometers attached to the whole hand

imal phalanges and the metacarpal area. The data were processed and analyzed using MATLAB (The MathWorks).

To investigate the effect of accelerometer attachment on skin dynamics, a preliminary test was performed. A 100 Hz sine stimulus applied to the tip of the finger, under two conditions: without accelerometer (Fig. 3.2a) and with an accelerometer attached at proximal phalanx (Fig. 3.2b). Ten trials were measured for each condition, and each lasted 4 s. For the latter condition, the accelerometer was detached and re-attached to the skin in each trial while the subject maintained the same posture. The laser Doppler vibrometer (LDV) measurements were synchronized. Signal waveforms of 100 ms measurements from both conditions are shown in Fig. 3.2c.

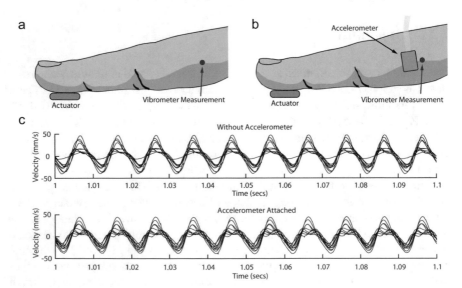

Fig. 3.2 LDV measurement of skin vibrations at proximal phalanx of index finger, (**a**) with and (**b**) without the accelerometer attached. For each case the measurement was repeated ten times (**c**)

Pearson's linear correlation coefficient score was computed for the ten pairs of signals for both conditions. The median correlation score = 0.91. The same score was computed for pairs of signals measured without the accelerometer (median = 0.91). Mann-Whitney U-test indicates that the correlation across both conditions and the correlation within the no-accelerometer condition are sampled from distributions with the same median (p = 0.40). The result of this preliminary test suggests that the attachment of accelerometer has minimal effect on skin dynamics.

3.2.2   Measurement Procedure

Sensor Placement

Our data acquisition (DAQ) system had 90 input channels, allowing in total 30 tri-axis accelerometers to be connected for measuring vibrations on the whole hand. We used fewer accelerometers in tasks where the measurement area on the hand is more localized. Four sensor configurations were used, with measurement coverage increased from a single digit to the whole hand:

1. One-finger configuration (15S), with nine accelerometers on the index finger and the remainder on the dorsal surface (Fig. 3.3a).
2. Two-finger configuration (15D), with six accelerometers on each of the first and second finger and three on proximal areas of the dorsal surface of the hand (Fig. 3.3b).
3. Four-finger configuration (15W), nine accelerometers (black disks) were placed on fingers 2, 3, and 4, with one accelerometer on each of the distal, medial, and proximal phalangeal sections, two on the distal and proximal phalangeal section of the thumb, and three on the dorsal surface (Fig. 3.3c).
4. Whole-hand configuration (30W), sensor placement on the fingers, and dorsal surface of the hand under 30 accelerometers configuration (Fig. 3.3d).

To ensure that the measurements were recorded in conditions that were consistent for all participants in the experiments, we positioned the sensors on each hand in configurations that were as identical as possible relative to the anatomy of each participant. We avoided positioning the sensors on joints, where the skin surface deforms most during digit movement. Instead, sensors are positioned on or between bones. This also reduced interference introduced by movement and ensured consistent attachment of the sensors. The positions were labeled anatomically as in Fig. 3.3. All participants were right-handed, and all measurements were recorded of the right hand. Sensors positioned on the digits were labeled with corresponding Roman numerals (I–V). The nearest bone to each sensor was indicated with a letter so that D represented distal phalange, M represented middle phalange, and P represented proximal phalange. Met represents metacarpus, and betwMet indicated the region between two metacarpal bones. Where two or more sensors were located near to the same bone, the last part of the label was used to indicate the relative position on that bone. U

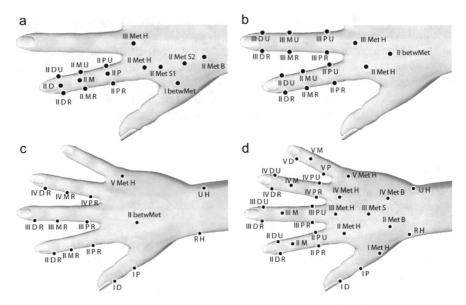

Fig. 3.3 Sensor placement on the fingers and dorsal surface of the hand. **a** One-finger configuration (15S). **b** Two-finger configuration (15D). **c** Four-finger configuration (15W). **d** Whole-hand configuration (30W)

represents ulna or toward the ulnar side of the sagittal plane, whereas R represents radius for the radial side. H referred to the head of the bone, and B referred to the base of the bone. S (or S1, S2) was used to indicate sensor locations on the shaft of the bone.

Measurement Protocol

We captured vibrations in the skin of the hand and fingers during a variety of manual interactions with different objects, materials, and parts of the hand (Table 3.1).

Contact Materials

We measured finger tapping and sliding interactions with object of different materials commonly encountered in our daily environment, including three materials with perceptually distinctive stiffness: steel, fabric, and hand skin, and three with perceptually distinctive roughness: steel, wood, and foam.

Table 3.1 Interaction modes and objects

Digits	Interaction	Object	Measurement
(II) (II, III)	Tap	Flat steel plate	15S 15D
(II) (II, III)	Tap	Fabric layer	15S 15D
(II) (II, III)	Tap	Dorsal hand skin	15S 15D
(II) (II, III)	Slide	Flat steel plate	15S 15D
(II) (II, III)	Slide	Wood surface	15S 15D
(II) (II, III)	Slide	Foam block	15S 15D
(I) (II) (III) (II, III) (all)	Light tap	Flat steel plate	15W
(I) (II) (III) (II, III) (all)	Hard tap	Flat steel plate	15W
(I) (II) (III) (II, III) (all)	Light slide	Flat steel plate	15W
(I) (II) (III) (II, III) (all)	Hard slide	Flat steel plate	15W
(I, II) (I, II, III) (all)	Precision grip	Glass cup	15W
(I, II) (I, II, III) (all)	Power grip	Glass cup	15W
(I, II) (I, II, III)	Indirect tap	Plastic stylus	15W
(I) (II) (III) (IV) (V)	Tap	Flat steel plate	30W
(II, III) (II, III, IV, V) (all)	Tap	Flat steel plate	30W
(II)	Slide	Flat steel plate	30W
(I, II)	Precision grasp	Small plastic cylinder	30W
(I, II)	Precision grasp	Large plastic cylinder	30W
(all)	Power grasp	Plastic ball	30W
(I, II)	Indirect tap	Plastic stylus	30W

Tapping and Sliding Force

The effect that interaction force has on cutaneous vibration patterns was also investigated. After being trained using a digital scale, our subjects performed tapping and sliding interactions with controlled light ($\approx 0.1\,\mathrm{N}$) and high force ($\approx 2.0\,\mathrm{N}$) on the top of a flat, steel plate. The measurement included interactions with number of digits involved, ranging from one to five. The entire dorsal surface of the hand was measured (15W).

Object Manipulation

Hand interactions with objects were measured, including precision grip (object contact with fingertips) and power grip (object contact with the palm and fingers) with a glass cup using two, three, or all digits, and indirect tapping by holding a plastic stylus with two or three digits.

The experiments were approved by the institutional research ethics review board of Drexel University. Informed consent was obtained in writing before the experiments. For experiment 1, two volunteers (male students at Drexel University, 22 and 23 years old, dominant right hand) wore the array of 15 accelerometers as indicated in Fig. 3.1. They sat in front of a table on which they rested their right forearms. In single and double finger measurements (15S and 15D), they performed tapping and sliding tasks on the specified surfaces. They performed other interactions in whole-hand (15W) measurements. Subjects were instructed to perform natural finger movements, and no restraint was applied to the inactive fingers, to avoid interfering with the movements. Each block of trials lasted 45 s and comprised 20 tapping trials and 10 trials for the other cases. Subjects were trained to follow a visual cue supplied by a computer to maintain a pace of 4 s per trial (2 s in the tapping condition). They performed the tasks with light (≈ 0.1 N) and high force (≈ 2.0 N). Grasping tasks involved precision and power grip of a glass cup and tapping using a plastic stylus. Precision grip is when the distal phalanges and the thumb tip press against each other on an object, whereas the power grip is when the fingers and palm clamp on an object with the thumb producing counter pressure. The sensor placement of experiment 1 is shown in Fig. 3.3a, b, and c (corresponding to 15S, 15D, and 15W, respectively).

In experiment 2, four volunteers (one female and three male students at the Drexel University, aged 19–23 years old, all right-hand dominant) wore the array of 30 accelerometers (Fig. 3.1). Measurement positions (Fig. 3.3d) were chosen to ensure that the accelerometers were evenly distributed, and the positions were standardized with respect to hand anatomy. Measurements were acquired as subjects performed specified actions with different parts of the hand and objects. Subjects performed tapping tasks 20 times and the other tasks 10 times. The gestures were selected to be similar to those used when interacting with the environment in everyday life. The majority involved coupled movement of multiple digits and contact between different parts of the hand and objects: tapping a steel plate with individual digits or combinations of digits, feeling the surface via sliding contact, two-finger grasping of a small or large plastic cylinder (diameter $d = 40$ or 56 mm, masses $m = 31$ g) with digits I and II, grasping a plastic ball ($d = 63$ mm, $m = 26$ g) with all fingers, and indirectly tapping a surface via a stylus ($d = 6$ mm, length $L = 155$ mm, $m = 30$ g) held in digits I and II. Subjects were instructed to use forces of approximately 1 N. The gestures were otherwise unconstrained.

Data Processing

To ensure that the frequency content of the measurements included the range of PC sensitivity [1], the data were minimally filtered. We used a zero-phase 10-Hz high-pass filter to eliminate the effects of hand kinematics. Each measurement recording lasted 45 s. For tapping tasks, each 2 s trial involved the finger contacting and sliding against the plate and then returning to its original position. For sliding tasks, a trial

Fig. 3.4 Cutaneous
wavelength of frequencies
within relevant perception
range. Estimated from the
data of cutaneous wave
propagation speed measured
in [21]

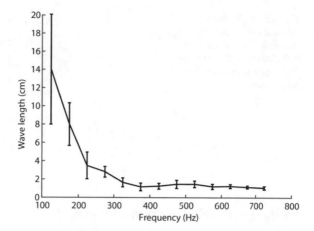

consisted of a combined forward and backward sliding motion 4 s in duration. For
gripping tasks, the fingers flexed and held the object during the first 2 s of the trial
and then extended and released the object (also 2 s).

We computed the acceleration magnitude $\mathbf{a}_i(t) = \sqrt{a_{i,x}(t)^2 + a_{i,y}(t)^2 + a_{i,z}(t)^2}$
for every recordings, where $a_{i,k}(t)$, $k = x, y, z$ are along the three-axis of the 3D
acceleration signals. The magnitude of the accelerometer measurement is independent of its mounting orientation on the hand, which allows comparison and fusion
of the data of all accelerometers. For each accelerometer i, the magnitude was
averaged over time by root mean square (RMS) to acquire the vibration energy
$\bar{\mathbf{a}}_i = \sqrt{\sum_t |\mathbf{a}_i(t)|^2}$.

3.2.3 Reconstructing Whole-Hand Signals from Sparse Measurement Locations

By interpolating the acceleration magnitude of all accelerometers using a physiologically based model of vibration propagation in the hand [21], we reconstructed the
vibration distribution over the entire hand.

The spatial wavelengths of signals excited in the range of frequencies relevant
to tactile perception are relatively long ($\lambda > 1$ cm, as shown in Fig. 3.4), allowing a
sparse Nyquist sampling approach. Both time-varying $\mathbf{a}_i(t)$ and RMS time-averaged
(A_i) skin vibrations across the hand can be estimated from accelerometer measurements, by mapping them onto a hand. Each accelerometer signal $\mathbf{a}_i(t)$ (or A_i) is
associated with a position $\mathbf{p}_i$. We associate these locations to the coordinates of the
3D hand model via anatomical registration of the anatomically known sensor locations. Using the model, we compute the geodesic distance $d(\mathbf{p}_i, \mathbf{p})$ between sensor
positions $\mathbf{p}_i$ and arbitrary surface points $\mathbf{p}$, using a shortest path algorithm. We estimated the acceleration magnitude $\mathbf{a}(\mathbf{p}, t)$ at arbitrary hand surface points using a

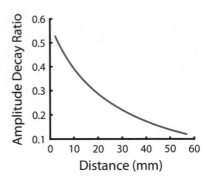

Fig. 3.5 A physiologically informed distance weighting function (Eq. 5.8) used to estimate the acceleration at arbitrary locations on the 3D hand model

physiologically informed distance weighting [21]:

$$\mathbf{a}(\mathbf{p}, t) = \frac{\sum_{i=1}^{30} f(\phi(\mathbf{p}, \mathbf{p}_i))\mathbf{a}_i(t)}{\sum_{i=1}^{30} f(\phi(\mathbf{p}, \mathbf{p}_i))}, \tag{3.1}$$

$$\phi(\mathbf{p}, \mathbf{p}_i) = \frac{\beta}{d(\mathbf{p}, \mathbf{p}_i) + \alpha} - C, \tag{3.2}$$

$$f(\phi) = \begin{cases} \phi, & \phi \geq 0 \\ 0, & \phi < 0. \end{cases} \tag{3.3}$$

Here, $f(\phi)$ is a rectifier (replacing all negative values with zeros). We evaluated these equations using values $\alpha = 25.5\,\text{mm}$, $\beta = 17.0\,\text{mm}$, and $C = 8.7 \times 10^{-2}$ that we selected based on previously published measurements [21]. $\phi(\mathbf{p}, \mathbf{p}_i)$ is used to compute the vibration amplitude at points $\mathbf{p}$ based on measurement of sensor $\mathbf{a}_i(t)$ at $\mathbf{p}_i$ accounting for damping with distance $d(\mathbf{p}_i, \mathbf{p})$ (Fig. 3.5). While we could readily accommodate hand size, via a scale parameter γ, differences in hand shape and mechanics would require further steps. For in this chapter and Chap. 4, skin vibration reconstructions were visualized using intensity maps rendered on a prototype hand, with deep blue corresponding to the minimum value and bright red to the maximum value.

3.3   Spatial and Time Domain Analysis

3.3.1   Spatial Patterns of Whole-Hand Vibrations

RMS intensity varied systematically over the dorsal surface of the hand (Figs. 3.6, 3.7, 3.8, and 3.9) and visibly depended on the contact interactions that produced them. In all cases, contact occurred near the distal end of the volar surface of the fingers, eliciting mechanical vibrations that propagated through the tissues of the hand. The

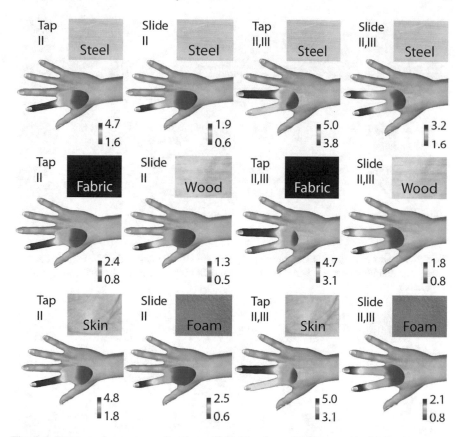

Fig. 3.6 Patterns of cutaneous vibrations elicited by fingertip contact with different materials. Surface palpation by tapping and sliding, in one-finger (1SS) and two-finger (1SD) sensor configurations. The amplitude range is normalized for each condition to enhance the distinguishability of the patterns

resulting patterns of vibration reflected the type of interaction, the locations of contact with the hand, the objects, and the materials involved.

As should be expected, the areas closest to the contact region were the most excited. Vibration intensity decayed with distance but could be easily detected by our apparatus far beyond the fingers, achieving maximum peak to peak amplitudes greater than $30\,\text{m/s}^2$ at all locations, in all conditions tested, which is well above perceptual and physiological thresholds [3, 23]. The different interaction modes gave rise to qualitatively distinct spatial distributions of intensity. The results also indicate that the range of vibration intensity appeared to be larger for contact interactions with hard objects than with very soft ones (Fig. 3.6). Moreover, there were systematic differences in vibration propagation patterns for different types of interaction. Tapping with multiple digits elicited broadly distributed patterns of vibration intensity, whereas sliding contact elicited more localized vibration (Fig. 3.7). Similarly, inter-

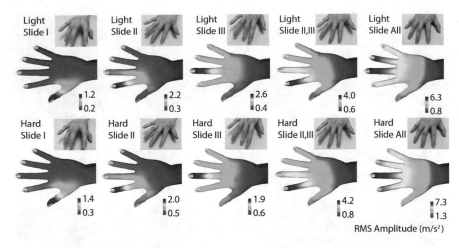

Fig. 3.7 Patterns of cutaneous vibrations elicited by sliding. Whole-hand (15 W) sensor configurations. The amplitude range is normalized for each condition to enhance the distinguishability of the patterns

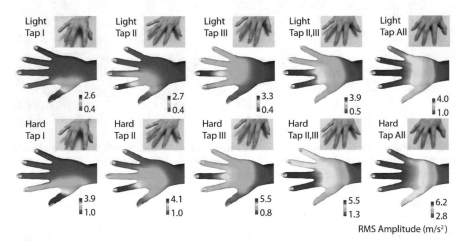

Fig. 3.8 Patterns of cutaneous vibrations elicited by tapping. Whole-hand (15 W) sensor configurations. The amplitude range is normalized for each condition to enhance the distinguishability of the patterns

actions at higher contact forces elicited more widely distributed patterns of vibration than lower forces did, even when normalized for intensity (Fig. 3.8).

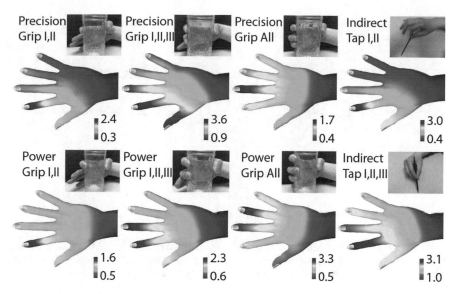

Fig. 3.9 Patterns of touch-elicited cutaneous vibrations. Surface palpation by whole-hand (15 W) sensor configurations. The vibrations were elicited by contact with different grasp types. The grasped objects consisted of a glass cup and a plastic stylus. The amplitude range is normalized for each condition to enhance the distinguishability of the patterns

3.3.2 Time Domain Correlates of Touch Interactions

Mechanical vibrations propagating in the skin also reflected the time course of inter-action between the hand and touched objects. For example, Fig. 3.10 illustrates the spatiotemporal pattern of vibration that was elicited when a participant tapped two digits (II, III) on a steel plate, as recorded from a single trial. Because motor behavior is much slower than vibration propagation, gross differences between spatial patterns at successive instants in this example could be attributed to contact timing rather than vibration propagation. Salient events, including asynchronous contact of digits (II) and (III) (delay 10 ms), are readily observed. Similar temporal patterns were observed in another hand interaction, involving movement of all digits (Fig. 3.11). Contact at the distal end of the digit yielded vibrations that propagated along the digit, across the dorsal surface, and to the wrist, before dissipating.

We observed rapid changes in the patterns of intensity over time, as cutaneous vibrations propagated unevenly on the dorsal side of the hand. Vibration energy peaked dramatically in time and space on the contact of a finger with an object and then spread quickly. Within a few milliseconds, its intensity reached a maximum, and then faded out within 20–30 ms. Owing to the impulsive nature of the stimulation, the signals that were observed were highly asymmetric in time.

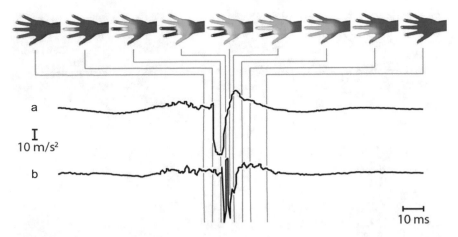

Fig. 3.10 Spatiotemporal distribution of vibration intensity from a single recording when tapping digits I and II on a steel plate (configuration 30 W). The time course of evolution of acceleration (y-axis) at locations on the distal phalangeal area of digit III **a** and digit II **b** are shown

3.4 Frequency Domain Analysis

To further characterize spatiotemporal variations in touch-induced vibrations, we constructed frequency-dependent portraits of RMS vibration intensity. We band-pass filtered the RMS acceleration signals to separate them into different frequency bands (0.1–10, 10–100, 100–200, 200–400, 400–700, and 700–1000 Hz) and constructed intensity maps for each, using the same method described in Sect. 3.2.3. The filter was realized by performing fast Fourier transform (FFT) on each component axis of the accelerometer signal and separating into distinct bands. Spatial vibration intensity distributions were computed for each frequency band. To compare the relative amplitude in each band, the same intensity scale was used for all frequency bands associated with a given gesture, but different gestures were normalized independently (Fig. 3.12). The lowest frequency band, from 0.1 to 10 Hz, included motor information because the typical timescale for finger movement was 2 s. The content below 0.1 Hz was largely due to gravity.

There were noticeable differences between direct and indirect tapping gestures. For gestures involving direct finger contact, vibration intensity in the frequency band 10–100 Hz always had the highest amplitude. For indirect tap with a stylus, however, vibration intensity peaked in the bands from 0.1 to 10 and 200–400 Hz. When performing indirect tap, vibration energy was transmitted to the digits that were not in contact with the stylus. This occurred primarily at low frequencies, 100 Hz. Vibrations produced via direct tap were lower in amplitude in higher frequency bands above 10–100 Hz. In contrast, those elicited by gripping a ball varied greatly among frequency bands. We also observed significant differences between patterns elicited by gripping a ball and by contacting a steel plate with the same fingers, especially

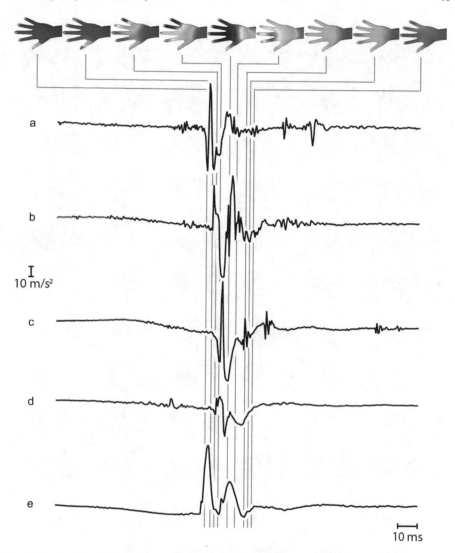

Fig. 3.11 An illustration of the spatiotemporal distribution of vibration intensity arising during a single interaction, consisting of tapping all digits on a steel plate. The y-axis acceleration recordings from the distal phalangeal area of each moving finger (sensor locations V D, IV D R, III D R, II D R, and I D)

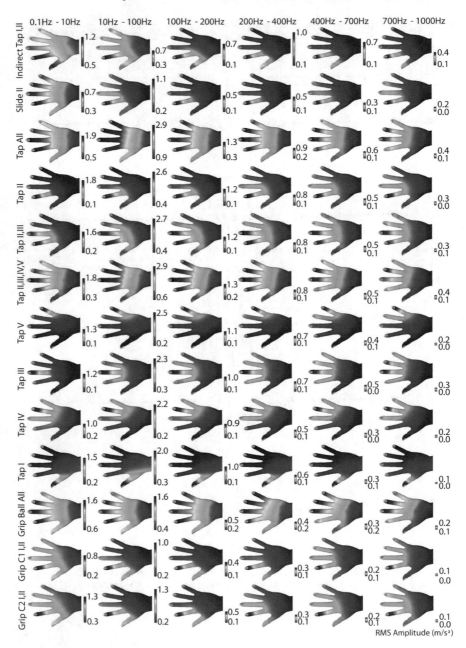

Fig. 3.12 Examples of spatial patterns of vibration intensity in six nonoverlapping frequency bands for 13 classes of gestures. Each subplot was normalized over itself

in the band of 10–100 Hz. In bands 100 Hz, vibration intensity was generally low when gripping a ball, but decreased rapidly with frequency when contacting a steel plate with all fingers.

3.5 Tactile Signals Encode the Modes of Interaction

3.5.1 Decoding Hand Interactions

We investigated the possibility of decoding the modes of interaction from these signals using machine learning methods (Fig. 3.13). We trained a support vector machine (SVM) classifier to predict the interaction mode that gave rise to vibration signals that were recorded using the 30-sensor (whole-hand) configuration. For each data trial, we used RMS time-averaged amplitude of all 30 accelerometers as input features and the corresponding gestures as the labels. We combined the data from all participants in random order and reported classification performance as the average of tenfold cross-validation.

To accurately classify the grasping interactions, we used a two-level classification hierarchy, with the three gripping gestures discriminated in the second classification

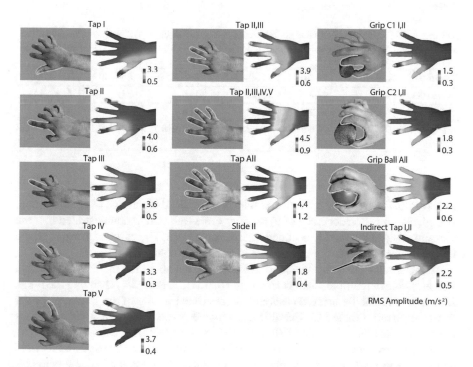

Fig. 3.13 Interaction modes and spatial patterns of vibration intensity, averaged between all four subjects (condition 30 W)

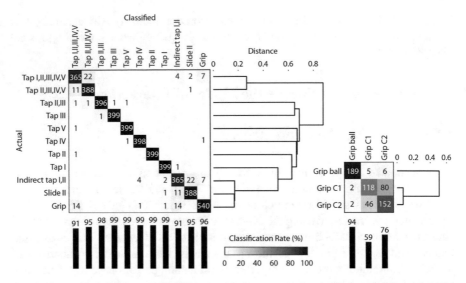

Fig. 3.14 Multi-class hierarchical SVM classification matrix for the 13 interaction modes. The second classification level disambiguates the grip type. The data from all four participants were combined for the analysis. Vertical bars report the percent correct for each class. The dendrograms, obtained from the MANOVA analysis, indicate the similarity (Mahalanobis distance) between class means

Table 3.2 Classification rates using six frequency bands for 11 classes of gestures. Tenfold cross-validation was repeated 20 times. Maximum (Max), minimum (Min), mean, and standard error (SE) of the rates are reported here

Band (Hz)	0.1–10	10–100	100–200	200–400	400–700	700–1000
Max (%)	96.7	97.4	96.5	96.5	93.0	92.2
Min (%)	93.3	95.2	91.7	92.4	90.0	86.3
Mean ± SE (%)	94.1 ± 0.1	96.4 ± 0.1	94.2 ± 0.1	94.1 ± 0.1	91.4 ± 0.1	89.7 ± 0.1

task (Fig. 3.14). We used a confusion matrix to report the patterns of classification. A high classification accuracy of 97%, after cross-validation, demonstrated that the vibration data alone readily encoded interaction modes. The cases of grasping large or small cylinders were typically the only ones to be confused.

Using similar methods, we found that information about the mode of interaction was available in multiple, distinct frequency bands (see Sect. 3.4) spanning the range salient to vibrotactile using measurements restricted to any of six nonoverlapping frequency bands (Table 3.2). Classification accuracy was greater than 89% in every band, with the highest rate (96.5%) achieved for the 10–100 Hz band.

Differences between the hands of individual subjects could be expected to yield differences in patterns of touch-elicited vibrations. To assess the between-subjects generalization of classification performance, we attempted to decode the data from

Table 3.3   MANOVA results of all configurations. dF: maximum dimension of group mean space. p: Maximum p-value for all dimensions. λ: Wilk's Lambda for the same dimension

Configuration	dF	p	λ	Most distinguishable pairs
15S	5	10^{-11}	0.50	Slide (wood), Tap (skin)
15D	5	10^{-9}	0.55	Slide (wood), Tap (skin)
15W	15	7×10^{-6}	0.90	Tap (I), Tap (I, II, III, IV, V)
30W	12	10^{-11}	0.98	Tap I, Grip cylinder

each participant using a classifier that we trained on data collected from the other three participants. The classification results were averaged over all participants and were computed via tenfold cross-validation. The resulting rates averaged 84.1%, indicating that despite individual differences, the information content in these signals was quite resilient, although the small size and relative homogeneity of the subject pool should be noted.

3.5.2   Statistical Differences of Tactile Signals

MANOVA was used to test for statistically significant differences among the gesture classes in the same dataset and to assess the pairwise distinguishability of different classes (Table 3.3). For all sensor configurations, vibration patterns were heterogeneous between interaction modes (MANOVA, $p < 10^{-5}$). The most distinguishable cases included sliding on wood vs. tapping the skin and grasping a cylinder vs. tapping a finger on a steel plate. Using similar methods, we found that information about the mode of interaction was available in multiple, distinct frequency bands spanning the range salient to vibrotactile perception.

3.6   Tactile Signals in Dorsal and Volar Hand Surfaces

Our main findings were derived using measurements captured in the dorsal surface of the back of the hand, rather than the volar (palmar) surface. We used this measurement configuration because it was necessary to leave the volar surface free during touch interactions with objects, in order to avoid contact with the sensors, which would produce artifacts.

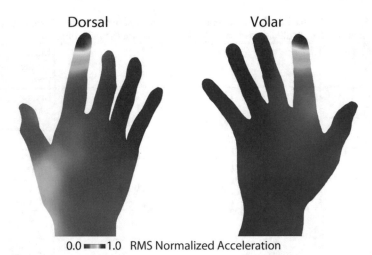

Fig. 3.15 The time-averaged tactile signals shown were produced via mechanical impulse stimulation with an electrodynamic exciter (Bruel & Kjaer, Model 4810) synchronized with a scanning laser Doppler vibrometer (Polytec, Model PSV-500), which sampled the acceleration of the skin at 400 discrete locations on each side of the hand. The fields shown are interpolated from the sampled acceleration data

We also captured measurements of vibrations in the palmar surface. These measurements show that stimulating of the hand at the areas used in our main experiments elicits similar tactile patterns in both the volar and dorsal hand surfaces. We captured these new data using two measurement techniques. The first was based on noncontact optical vibrometry measurements with a scanning laser Doppler vibrometer (model 8330, Ometron Ltd.). The second utilized a pair of wearable sensor arrays that were developed in our group [30]. The results (Figs. 3.15 and 3.16) reveal that when the hand is tapped at locations near those used in our main experiments, similar tactile patterns occur in both the volar and dorsal hand surfaces (Figure S8; methods described in caption). This suggests that even though our measurements were captured on the dorsal hand surface, very similar results would be obtained with the volar region of the hand. This finding is consistent with the physics of vibration transmission in the hand. In the physical regime relevant to tactile perception, the wavelengths of propagating vibrations are large, approximately 1–10 cm (Sect. 2.3), compared with many anatomical features. Such features are therefore expected to have a limited effect on vibration transmission in the hand. Furthermore, differences in the mechanical properties of hairy skin (which covers the dorsal hand surface) and glabrous skin (in the volar surface) are modest in the physical regime relevant to our results, for which the vibration displacements were typically smaller than 1 mm. The mechanical differences become significant for larger displacements. Finally, the volar and dorsal hand surfaces are coupled through tissue spanning the thickness of the hand. Thus, while we measured the analyzed stimuli at locations on the surface

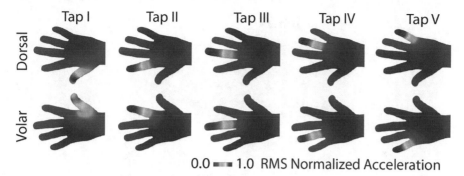

Fig. 3.16 Similar tactile signal patterns were also elicited in both hand surfaces via passive tapping at the tips of individual digits. Tapping was supplied by the electrodynamic exciter. A pair of custom arrays of 84 three-axis accelerometers were worn on the hand, one on each side

of the hand, such vibrations readily travel through sub-surface tissues that connect both sides of the hand, yielding similar vibration patterns in each hand surface, as reflected in our measurements.

3.7 Contact and Noncontact Elicited Vibrations

During normal hand function, vibrations produced via muscle and tendon activity are present, in addition to those elicited through skin–object contact. However, tendon- and muscle-produced vibrations are typically far smaller than those produced through

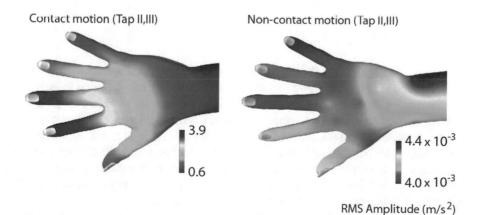

Fig. 3.17 Examples illustrating the difference in vibration acceleration magnitude at the skin during hand gestures with and without touch contact. Vibration acceleration magnitude was nearly three orders of magnitude smaller in the noncontact condition ($\approx 4 \times 10^{-3}$ m/s^2)

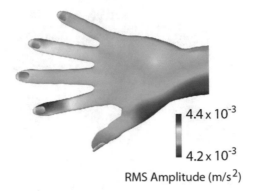

RMS Amplitude (m/s^2)

Fig. 3.18 The noncontact data extracted from the original data recording of all gestures (configuration 30 W). They were obtained by deleting a segment of time during each trial that corresponded to the touch contact phase and its immediate aftereffects. Because touch-elicited vibrations decayed rapidly, the deleted segments averaged less than 100 ms in duration

touch contact. To illustrate this, we processed our dataset (configuration 30W, all gestures) to remove time segments during and immediately after the touch contact phase of the manual interaction. Because touch-elicited vibrations decayed rapidly, the deleted segments averaged less than 100 ms in duration. The resulting accelerations in the noncontact condition proved to be nearly three orders of magnitude smaller than in the contact condition (Figs. 3.17 and 3.18). Although it could be argued that some of the movement-produced vibrations were excluded as a result of this processing, the results suggest that skin–object contact was the dominant source of the mechanical vibrations studied here.

3.8 Discussion

Cutaneous patterns of vibration vary in structured ways with the mode of interaction with a touched object, and these results demonstrate that it is possible to decode the interaction types directly from the vibration patterns they elicit. The classification analysis indicated that vibration patterns produced by tapping contact are highly distinctive from those produced by other gestures. In contrast, sliding contact, indirect tapping, and gripping gestures yielded similar multi-digit vibration distributions. Unsurprisingly, higher finger forces generally yielded higher vibration intensity, but also proportionally larger distances of propagation. Intensity and distance also increased with the number of digits engaged.

We also observed rapid changes in the patterns of intensity over time, as cutaneous vibrations propagated unevenly on the dorsal side of the hand. Vibration energy peaked dramatically in time and space on the contact of a finger with an object and then spread quickly. Within a few milliseconds, its intensity reached a maximum, and then faded out within 20–30 ms. Owing to the impulsive nature of the stimulation, the signals that were observed were highly asymmetric in time. Curiously, digit I, the thumb, produced lower intensities than the other digits.

Different manual gestures were observed to elicit distinct patterns of energy in the frequency domain, with indirect tapping yielding vibration energy that was concentrated at higher frequencies (between 200 and 400 Hz) than was the case for direct tapping (between 10 and 100 Hz), and these differences were preserved at locations distant from the areas of contact. Soft objects, such as the ball used in the grasping measurements, induced little energy 10 Hz. Thus, the mechanical characteristics of the contact affected the frequency content of the propagated energy. There was generally less energy in low-frequency bands. These, nonetheless, contained significant information, albeit within limits, because the kinematic differences elicited by variations in the size of a gripped object were lost.

Prior research has shed light on certain sensory specializations in the upper limb, including the high innervation density of the finger pads and the restriction of Meissner's corpuscles to glabrous skin, and their relevance to fine manual control. However, less is known about why some sensory cells, including PC units, are distributed more widely in the hand. The patterns of touch-elicited vibration, and the extent to which they can encode information about their source, may offer some explanation. The signals observed in this study have the greatest energy and spatial resolution in the fingers, although energies in the rest of the hand remained easily detectable by our apparatus. The large spatial scale of the variations in these patterns (on the order of 1 cm), and their fast temporal evolution (order 5 ms), could suggest that a sparse distribution of vibration-sensitive mechanoreceptors, similar to the network of PC units in the hand, would be appropriate to capture them. The proximity of extensor tendons at the dorsal surface of the hand suggests that muscle spindle afferents could play a role in processing vibrations during active touch [4, 24, 27], but more research is needed.

Further advances in our understanding of sensorimotor function in the upper limb may lead to new developments in prosthetic and robotic hands and to new technologies for providing realistic tactile feedback in virtual reality.

References

1. Bell, J., Bolanowski, S., Holmes, M.H.: The structure and function of pacinian corpuscles: a review. Prog. Neurobiol. **42**(1), 79–128 (1994)
2. Bicchi, A., Scilingo, E.P., De Rossi, D.: Haptic discrimination of softness in teleoperation: the role of the contact area spread rate. IEEE Trans. Robot. Autom. **16**(5), 496–504 (2000)
3. Brisben, A., Hsiao, S., Johnson, K.: Detection of vibration transmitted through an object grasped in the hand. J. Neurophysiol. **81**(4), 1548–1558 (1999)
4. Burke, D., Hagbarth, K.E., Löfstedt, L., Wallin, B.G.: The responses of human muscle spindle endings to vibration of non-contracting muscles. J. Physiol. **261**(3), 673–693 (1976)
5. Cauna, N., Mannan, G.: The structure of human digital pacinian corpuscles (corpuscula lamellosa) and its functional significance. J. Anat. **92**(Pt 1), 1 (1958)

6. Cholewiak, R.W., McGrath, C.: Vibrotactile targeting in multimodal systems: accuracy and interaction. In: 2006 14th Symposium on Haptic Interfaces for Virtual Environment and Tele-operator Systems, pp. 413–420. IEEE (2006)

7. Delhaye, B., Hayward, V., Lefèvre, P., Thonnard, J.L.: Texture-induced vibrations in the forearm during tactile exploration. Front. Behav. Neurosci. **6**, 37 (2012)

8. Delhaye, B., Lefevre, P., Thonnard, J.L.: Dynamics of fingertip contact during the onset of tangential slip. J. R. Soc. Interface **11**(100), 20140698 (2014)

9. Edin, B.B., Abbs, J.H.: Finger movement responses of cutaneous mechanoreceptors in the dorsal skin of the human hand. J. Neurophysiol. **65**(3), 657–670 (1991)

10. Goodwin, A., Youl, B., Zimmerman, N.: Single quickly adapting mechanoreceptive afferents innervating monkey glabrous skin: response to two vibrating probes. J. Neurophysiol. **45**(2), 227–242 (1981)

11. Gottschaldt, K.M., Vahle-Hinz, C.: Merkel cell receptors: structure and transducer function. Science **214**(4517), 183–186 (1981)

12. Järvilehto, T., Hämäläinen, H., Soininen, K.: Peripheral neural basis of tactile sensations in man: ii. characteristics of human mechanoreceptors in the hairy skin and correlations of their activity with tactile sensations. Brain Res. **219**(1), 13–27 (1981)

13. Johansson, R.S., Flanagan, J.R.: Coding and use of tactile signals from the fingertips in object manipulation tasks. Nat. Rev. Neurosci. **10**(5), 345–359 (2009)

14. Johansson, R.S., Landstro, U., Lundstro, R., et al.: Responses of mechanoreceptive afferent units in the glabrous skin of the human hand to sinusoidal skin displacements. Brain Res. **244**(1), 17–25 (1982)

15. Johansson, R.S., Vallbo, A.: Tactile sensibility in the human hand: relative and absolute densities of four types of mechanoreceptive units in glabrous skin. J. Physiol. **286**(1), 283–300 (1979)

16. Jörntell, H., Bengtsson, F., Geborek, P., Spanne, A., Terekhov, A.V., Hayward, V.: Segregation of tactile input features in neurons of the cuneate nucleus. Neuron **83**(6), 1444–1452 (2014)

17. Kumamoto, K., Senuma, H., Ebara, S., Matsuura, T.: Distribution of pacinian corpuscles in the hand of the monkey, macaca fuscata. J. Anat. **183**(Pt 1), 149 (1993)

18. Libouton, X., Barbier, O., Berger, Y., Plaghki, L., Thonnard, J.L.: Tactile roughness discrimination of the finger pad relies primarily on vibration sensitive afferents not necessarily located in the hand. Behav. Brain Res. **229**(1), 273–279 (2012)

19. Mahns, D.A., Perkins, N., Sahai, V., Robinson, L., Rowe, M.: Vibrotactile frequency discrimination in human hairy skin. J. Neurophysiol. **95**(3), 1442–1450 (2006)

20. Maksimovic, S., Nakatani, M., Baba, Y., Nelson, A.M., Marshall, K.L., Wellnitz, S.A., Firozi, P., Woo, S.H., Ranade, S., Patapoutian, A., et al.: Epidermal merkel cells are mechanosensory cells that tune mammalian touch receptors. Nature **509**(7502), 617–621 (2014)

21. Manfredi, L.R., Baker, A.T., Elias, D.O., Dammann III, J.F., Zielinski, M.C., Polashock, V.S., Bensmaia, S.J.: The effect of surface wave propagation on neural responses to vibration in primate glabrous skin. PloS One **7**(2) (2012)

22. Manfredi, L.R., Saal, H.P., Brown, K.J., Zielinski, M.C., Dammann, J.F., III., Polashock, V.S., Bensmaia, S.J.: Natural scenes in tactile texture. J. Neurophysiol. **111**(9), 1792–1802 (2014)

23. Merzenich, M.M., Harrington, T.: The sense of flutter-vibration evoked by stimulation of the hairy skin of primates: comparison of human sensory capacity with the responses of mechanore-ceptive afferents innervating the hairy skin of monkeys. Exp. Brain Res. **9**(3), 236–260 (1969)

24. Ribot-Ciscar, E., Rossi-Durand, C., Roll, J.P.: Muscle spindle activity following muscle tendon vibration in man. Neurosci. Lett. **258**(3), 147–150 (1998)

25. Robles-De-La-Torre, G., Hayward, V.: Force can overcome object geometry in the perception of shape through active touch. Nature **412**(6845), 445–448 (2001)

26. Rogers, C.H.: Choice of stimulator frequency for tactile arrays. IEEE Trans. Man-Mach. Syst. **11**(1), 5–11 (1970)

27. Roll, J., Vedel, J.: Kinaesthetic role of muscle afferents in man, studied by tendon vibration and microneurography. Exp. Brain Res. **47**(2), 177–190 (1982)

28. Saal, H.P., Bensmaia, S.J.: Touch is a team effort: interplay of submodalities in cutaneous sensibility. Trends Neurosci. **37**(12), 689–697 (2014)

29. Shao, Y., Hayward, V., Visell, Y.: Spatial patterns of cutaneous vibration during whole-hand haptic interactions. Proc. Natl. Acad. Sci. **113**(15), 4188–4193 (2016)
30. Shao, Y., Hu, H., Visell, Y.: A wearable tactile sensor array for large area remote vibration sensing in the hand. IEEE Sens. J. **20**(12), 6612–6623 (2020)
31. Sofia, K.O., Jones, L.: Mechanical and psychophysical studies of surface wave propagation during vibrotactile stimulation. IEEE Trans. Haptics **6**(3), 320–329 (2013)
32. Stark, B., Carlstedt, T., Hallin, R., Risling, M.: Distribution of human pacinian corpuscles in the hand: a cadaver study. J. Hand Surg. **23**(3), 370–372 (1998)
33. Tanaka, Y., Horita, Y., Sano, A.: Finger-mounted skin vibration sensor for active touch. In: International Conference on Human Haptic Sensing and Touch Enabled Computer Applications, pp. 169–174. Springer (2012)
34. Visell, Y., Giordano, B.L., Millet, G., Cooperstock, J.R.: Vibration influences haptic perception of surface compliance during walking. PLoS One **6**(3), e17697 (2011)
35. Wiertlewski, M., Hayward, V.: Mechanical behavior of the fingertip in the range of frequencies and displacements relevant to touch. J. Biomech. **45**(11), 1869–1874 (2012)
36. Wijntjes, M.W., Sato, A., Hayward, V., Kappers, A.M.: Local surface orientation dominates haptic curvature discrimination. IEEE Trans. Haptics **2**(2), 94–102 (2009)

Chapter 4
Compression of Dynamic Tactile Information in the Human Hand

Abstract Chapter 3 provided a new view of human tactile sensing, revealing how the transmission of mechanical wave signals in the hand transforms localized touch contact into a variety of spatiotemporal vibration patterns. To better understand how sense organs extract perceptual information from the physics of the environment, this chapter investigates the information content in these signals via a data-driven optimal encoding framework based on convolutional non-negative matrix factorization. It shows that these signals produce an efficient encoding of tactile information. The computation of an optimal encoding of thousands of naturally occurring tactile stimuli yielded a compact lexicon of primitive wave patterns that sparsely represented the entire dataset, enabling touch interactions to be classified with an accuracy exceeding 95%. The primitive tactile patterns reflected the interplay of hand anatomy with wave physics. Notably, similar patterns emerged when we applied efficient encoding criteria to spiking data from populations of simulated tactile afferents. This finding suggests that the biomechanics of the hand enables efficient perceptual processing by effecting a preneuronal compression of tactile information. To clarify the results, a chapter appendix also presents a brief analysis of the mathematics of elastic wave transmission in the skin. This analysis led to the wave-based tactile feedback methods that are presented in Chap. 6.

Disclaimer

© The Author(s), under exclusive license to Springer Nature Switzerland AG 2022   53
Y. Shao, *Tactile Sensing, Information, and Feedback via Wave Propagation*,
Springer Series on Touch and Haptic Systems,
https://doi.org/10.1007/978-3-030-90839-3_4

4.1   Introduction

The sense of touch, which is essential for skilled manipulation and object perception, relies on the encoding of mechanical signals collected by the skin and subcutaneous tissues into neural representations. While neural responses to tactile stimuli are often associated with mechanical inputs arising from small skin regions, we recently observed that dynamic touch elicits mechanical waves in the tactile frequency range that spread throughout the whole hand, with transient excitations decaying within 30 ms [54]. An appendix to this chapter reviews salient aspects of the physics of mechanical waves in skin and other tissues in the physical regime relevant to tactile sensation and perception.

Dynamic tactile inputs can thus drive widespread tactile afferent populations [38, 47]. These touch-elicited waves have been found to facilitate fine perceptual discriminations [35, 41], and can be used to infer actions, the attributes of touched objects, and locations of contact with the hand [14, 39, 54, 65]. In fact, receptive fields of neurons in somatosensory cortical areas were observed to span large hand areas and multiple digits [23, 59]. The large spatial-scale integration at the early stages of processing [8, 25] induces cortical neurons to exhibit integrative responses to tactile inputs delivered to widespread limb regions [15, 19, 46]. Thus, somatosensory processing could depend on information transported by mechanical waves that propagate in tissues to remote locations, distant from the loci of mechanical contact.

An analogy could be drawn to the cochlea, where the transport of dispersive mechanical waves via the basilar membrane imparts preneuronal filtering to auditory stimuli [43], supporting a frequency-place transformation [10, 13, 18, 63]. Similar processes have been observed for mechanical waves propagating in the hand [38]. In the rodent vibrissal system, whisker mechanics also impart preneuronal processing to tactile stimuli [3, 42].

If the transport of mechanical waves in the hand facilitates efficient somatosensory information encoding, it should be possible to describe tactile stimuli in terms of a smaller space of informative parameters. This would allow stimuli to be represented as combinations of a small number of primitive features or tactile patterns. Such representations are commonly observed in sensory systems. They correspond to an efficient sensory coding hypothesis that proposes neural circuitry to have evolved to capture relevant sensory information with the fewest physical and metabolic resources [1, 4]. Studies of commonly encountered visual and auditory stimuli show that representations in the neural pathways for perceptual processing can emerge from the need to efficiently encode information in natural scenes [7, 12, 33, 44].

In this chapter, we show how mechanical waves in the hand produce an efficient encoding of tactile inputs. By optimally encoding a dataset of thousands of naturally occurring whole-hand tactile stimuli, we obtained a compact lexicon of primitive spatiotemporal patterns that sparsely represented information in the entire dataset, enabling it to be classified with an accuracy exceeding 95%. These primitive patterns reflected the interplay of the anatomy of the hand and the physics of tactile wave propagation and were evocative of hand sensory function, including the individuation of digits and the denser innervation of the distal ends of the fingers. We obtained strikingly similar patterns when we applied the efficient encoding criteria to spiking data from populations of simulated tactile afferents. These results reveal a possible important contribution of the hand biomechanics to early somatosensory processing, which may be compared to the role of cochlear mechanics in early auditory encoding. This new knowledge revises existing views of touch sensing and may aid the understanding of hand sensory function and deficits affecting the sense of touch. It also furnishes new principles that may guide the design of electronic tactile sensors that could leverage the ability of propagating waves to communicate touch information. Such devices may yield important applications in robotics, prosthetics, and medicine.

4.2 A Data-Driven Model of Efficient Tactile Encoding

We formulated the efficient encoding of tactile information as an optimal matrix factorization problem and evaluated its predictions using the database of whole-hand tactile signals acquired in Chap. 3.

The data comprised spatiotemporal skin accelerations magnitude, $a(\mathbf{x}, t)$, which were captured at 30 different locations, $\mathbf{x}$, via a sensor array worn on the hand during performances of 13 manual gestures and 4,600 interactions with objects (Fig. 3.13). We measured from each interaction a duration of 600 ms of signals. For analysis, the signals were downsampled to 1.0 from 2.0 kHz. Still, for each interaction, 30 time-varying signals of 600 samples yielded in total 18,000 data samples. Thus, the nominal dimensionality of each of the 4,600 spatiotemporal stimuli in the dataset was 18,000.

In the following sections, we first encode the spatial tactile information using the RMS time-averaged measurements (see Sect. 3.2.2, Data processing). We encouraged sparseness of the encoding by adding constraints to the optimization. Then, spatiotemporal tactile information was encoded with a convolutional model that capture the time-invariant temporal structure in the tactile signals.

4.3 Spatial Encoding

It was found that non-negative matrix factorization (NMF) can discover latent factors in an unsupervised setting. This technique has found applications in many areas of engineering, biology, and the physical sciences [37, 57, 64]. Non-negativity implies that the model is additive, allowing the latent factors to be interpreted as parts-based summaries of the data [31].

Inspired by the idea that neural representations tend to be parsimonious [16], we applied NMF to infer the latent factors in the recorded tactile signals in the form of time-averaged signals, $\bar{a}_k$. If the time-averaged dataset is represented by a $I \times J$

matrix, V, where I is the number of channels and J is the number of recorded trials, NMF sought to approximate V by the product of a $I \times M$ basis matrix, W, with a $M \times J$ matrix of activation weights, H, both of which are constrained to be non-negative.

The constant, M, is the rank of the decomposition, that is, the number of bases. The rank is usually selected to be smaller than I and J. The approximation was realized by minimizing an objective function, D, given by the root mean square residual, $D = \|V - WH\|_F$, where $\| \cdot \|_F$ denotes the Frobenius matrix norm. The NMF representation could be expressed in explicit form. For each trial, the factorization approximated the data, $a(\mathbf{x})$, by $\hat{a}(\mathbf{x})$, where $\mathbf{x}$ was a point of measurement, up to a residual noise term η,

$$\hat{a}(\mathbf{x}) = \sum_{i=1}^{M} h_i \, w_i(\mathbf{x}), \qquad w_i(\mathbf{x}) \geq 0, \ \ h_i \geq 0, \forall i \, ,$$

$$\hat{a}(\mathbf{x}) = a(\mathbf{x}) + \eta \, .$$

The resulting basis patterns were evocative of hand anatomy, sensory innervation patterns, and contact regions. Several basis patterns were concentrated at the distal ends of each digit, with other fields spanning different hand regions (Fig. 4.1). Similar

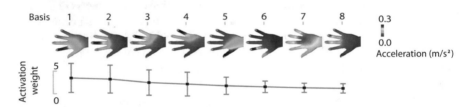

Fig. 4.1 Learned representations of time-averaged tactile waves evoke hand anatomy and sensory function. The bases are ordered by their mean activation weights toward encoding the hand interactions

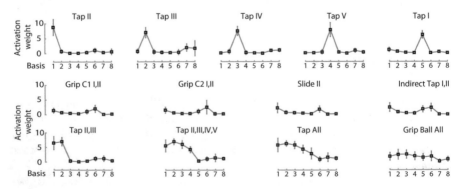

Fig. 4.2 Mean activation weights for each manual action with respect to each basis (Fig. 4.1) that encodes time-averaged tactile waves

motifs emerged when the rank N was varied. The corresponding activation weights varied systematically with the manual actions that produced them (Fig. 4.2), including the distribution of tactile stimuli in hand regions that were involved in the actions.

4.4   Spatiotemporal Encoding

Tactile sensing is a spatiotemporal process. Our encoding model of time-averaged tactile stimuli yielded primitive representations that reflected salient features of the anatomy and the function of the hand, but this model did not capture the temporal evolution of the tactile stimuli, which should be an essential component of any representation of tactile inputs.

A variety of models could be used to empirically investigate the encoding of spatiotemporal data, including variants of artificial neural network and deep learning models [5, 28, 29, 48], and state-space models [66], among others. Many of these methods require numerous arbitrary choices to be made about model structures and free parameters, obfuscating the interpretation of the results.

Other approaches to modeling temporal structure involve block-based segmentations of data that employ overlapped short-term windows of analysis within which signals are assumed to be stationary. Examples of such approaches are found in methods based on the short-time Fourier transform. The representations that result from these approaches can be inconsistent in time, can suffer windowing artifacts, can be inefficient, and can exhibit a high degree of entropy.

Here, we formulated the encoding as a linear, convolutional non-negative matrix factorization (CNMF), which is an extension of the approach used for the time-independent coding analyzed in the preceding section. Spatiotemporal stimuli were encoded via a compact lexicon of M primitive spatiotemporal patterns (bases), $w_i(\mathbf{x}, t)$, weighted by time-dependent activations, $h_i(t)$, that were unique to each stimulus (Eq. 4.2). The model included a non-negativity constraint to match the rectifying property of mechano-transduction [26]. No other assumptions were made about the statistics of the stimuli.

4.4.1   Mathematical Model and Optimization

These stimuli were encoded via a compact lexicon of M bases of length T, $w_i(\mathbf{x}, t)$, weighted by time-dependent activations, $h_i(t)$, that were unique to each stimulus.

$$\hat{a}(\mathbf{x}, t) = \sum_{i=1}^{M} \sum_{\tau=0}^{T-1} h_i(t - \tau) \, w_i(\mathbf{x}, \tau), \quad \text{where } w_i(\mathbf{x}, t) \geq 0, \quad h_i(t) \geq 0. \quad (4.1)$$

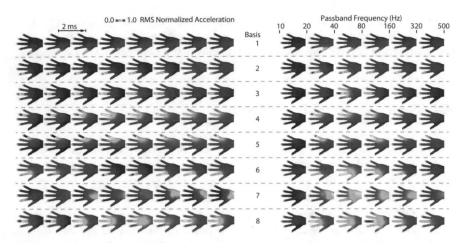

Fig. 4.3 Left: Each row represents a spatiotemporal basis pattern. The tactile bases, displayed at 2 ms intervals in descending order of activation, reflect the individuation of digits and larger representations of commonly used digits II and III. Right: Analysis within different frequency bands revealed that different basis patterns captured distinct frequency content

The model of efficient spatiotemporal encoding is based on CNMF [56]. This model may be compared to that used to represent stimulus encoding in the auditory system [34]. The model is mathematically simple, requires few arbitrary choices, can accommodate physiologically motivated assumptions, and can be compared with models of sensory encoding in other modalities. It encoded the tactile stimuli, $\mathbf{a}(\mathbf{x}, t)$, by determining the values of h_i and w_i that provided the best statistical estimate, $\hat{\mathbf{a}}(\mathbf{x}, t) = \mathbf{a}(\mathbf{x}, t) + \eta(\mathbf{x}, t)$, of $\mathbf{a}(\mathbf{x}, t)$ as determined by the model (4.2), where η is a residual error.

We computed an optimal encoding by maximizing the information about every element of the dataset, $\mathbf{a}(\mathbf{x}, t)$, that was gained by observing the estimate, $\hat{\mathbf{a}}(\mathbf{x}, t)$, as determined by the model (4.2). The simultaneous optimization of the model with respect to $w_i(\mathbf{x}, t)$ and $h_i(t)$ yielded a set of "tactile basis patterns", $w_i(\mathbf{x}, t)$ that together produced an efficient encoding, revealing the latent structure hidden in the ensemble of stimuli (an example of M = 8 is shown in Fig. 4.3). These basis patterns, $w_i(\mathbf{x}, t)$, optimally represented all 18000 sample values of all 4600 stimuli in the dataset in the sense of maximum likelihood. Owing to its time-dependence, the problem possessed more than 100-fold more degrees of freedom than the time-averaged case analyzed in the preceding section (Sect. 4.3). Nonetheless, the optimal spatiotemporal basis patterns were also evocative of hand anatomy and function, echoing our findings in the time-independent case.

Although the analysis was blind to the conditions that gave rise to the signals, the tactile bases were evocative of hand sensory function (Fig. 4.3). Most were initially localized at the distal ends of single digits (the most densely innervated regions of the hand). They traveled proximally at rates of 1–10 m/s, while decaying over 10–30 ms, matching the causal physics of waves in the hand (Sect. 3.3.2). Other bases evolved

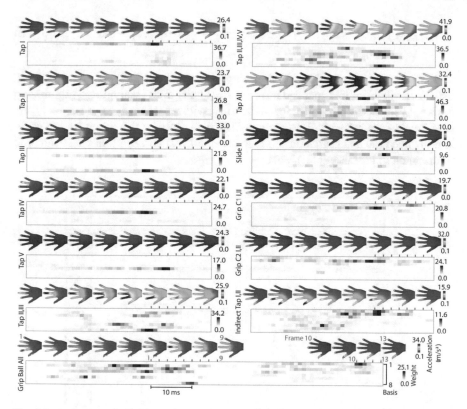

Fig. 4.4 Activations, $h_i(t)$ (shown in grayscale), produced by encoding the displayed tactile stimulus. Temporal resolution: 1 ms. Blue ticks show the time instant for each displayed stimulus frame

from the distal region of individual digits to diffuse regions of the hand surface. In the frequency domain, pairs of bases exhibited similar spatial patterns but distinct frequency characteristics. For example, the encoding yielded pairs of bases that were both spatially localized within one digit but possessed different filtering properties: low-pass, from about 20 to 80 Hz (Fig. 4.3, basis 2), or high-pass, from 80 to 160 Hz (Fig. 4.3, basis 6).

The activation weights, $h_i(t)$, associated with each basis differed for each stimulus (Fig. 4.4). Both factors were jointly estimated from the data. Each basis could assume arbitrary non-negative values for each position and time. No other statistical assumption was made about the data. The model was causal (4.2), hence the bases described responses that ensued with delays, τ. This optimization maximized the statistical information about the stimulus, $a(\mathbf{x}, t)$, that is gained by observing the estimate, $\hat{a}(\mathbf{x}, t)$ (4.2), as measured by the Kullback-Leibler (KL) divergence:

$$D_{\mathrm{KL}}(a(\mathbf{x}, t), \hat{a}(\mathbf{x}, t)) = \sum_{\mathbf{x}, t} \left(a(\mathbf{x}, t) \ln \frac{a(\mathbf{x}, t)}{\hat{a}(\mathbf{x}, t)} - a(\mathbf{x}, t) + \hat{a}(\mathbf{x}, t) \right). \qquad (4.2)$$

This measure quantified the dissimilarity between a and $\hat{a}$, regarding them as statistical distributions that encoded information. Under mild assumptions, minimizing the KL divergence is equivalent to maximizing, with respect to $h_i(t)$ and $w_i(\mathbf{x}, t)$, the likelihood of the model (4.2) to represent the data. Solving this minimization problem involved the determination of parameters $w_i(\mathbf{x}, t)$ and $h_i(t)$ that best captured information in the ensemble of data.

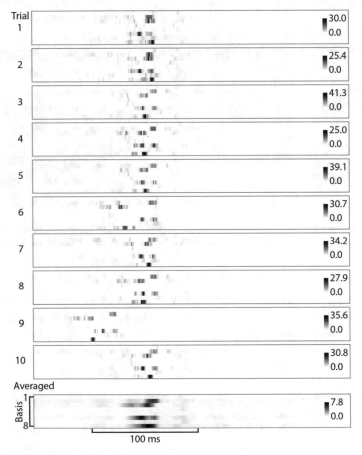

Fig. 4.5 Activation patterns for individual trials, and mean activations across all trials, for tactile stimuli associated with simultaneously tapping digits II and III. Individual trials elicited sparse activation patterns, which became blurred when averaged in time, due primarily to differences in the timing of finger-surface contact

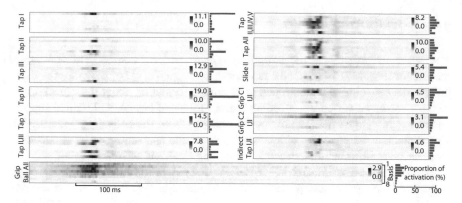

Fig. 4.6 Mean activations for stimuli elicited by each gesture class, averaged across all encoded trials

4.4.2 Activation Weight Patterns

The bases encoded the stimuli via a small number of time-dependent activation weights. Stimuli elicited by multi-finger gestures were encoded by several bases, while simpler gestures activated one or two. Tactile stimuli produced via similar gestures yielded similar activation patterns (Fig. 4.5). Therefore, activation weights of similar gesture were averaged to reveal the statistics of the activations of primitives (Fig. 4.6). In contrast, dissimilar gestures resulted in dissimilar activations, even when the same combinations of digits were involved.

4.4.3 Hyperparameter Optimization of the Encoding Model

We computed optimal encodings with ranks $M = 2$–12, corresponding to 2–12 basis patterns. For each value of M, we determined the optimal basis set and per-stimulus activation weights via simultaneous iterative optimization over w_i and h_i, beginning from random initializations of each [32]. We set their duration, T, to the time required for mechanical waves in the hand to decay, about 30 ms, as uncovered in Chap. 3 (Sect. 3.3.2). This duration spanned 30 time samples at the sample rate of 1 kHz. Each basis pattern was therefore represented by 900 values. Our findings were robust to variations in duration, as will be discussed in the next section.

4.5 Encoding Efficiency

The space of possible tactile stimuli is constrained by contact and continuum mechanics (Fig. 4.7). To assess the number of bases, $w_i(\mathbf{x}, t)$ (or $w_i(\mathbf{x})$ for the time-independent case), that were needed to capture information about the causal origin of the stimuli, we varied the encoding rank (number of bases), and trained support vector machine classifiers to predict the gestures from the activation patterns. We also evaluated the encoding efficiency based on the sparseness of each encoding rank, showing information independence among the basis activations.

We designed a classification task whose objective was to use the activation weight pattern to identify the gesture that elicited the stimulus. We integrated the weights over time, $\bar{h}_i = \frac{1}{T} \sum_{t=0}^{T-1} h_i(t)$, to eliminate the adverse effects arising from timing differences across trials. The task involved multi-class classification, which we implemented as thirteen 1-versus-12 classification tasks. We avoided classification methods, such as convolutional neural networks, that would require extensive model tuning. We instead opted for SVM classifiers, which require few choices and are

0.1 ▬▬▬ 3.0 RMS Acceleration(m/s²)

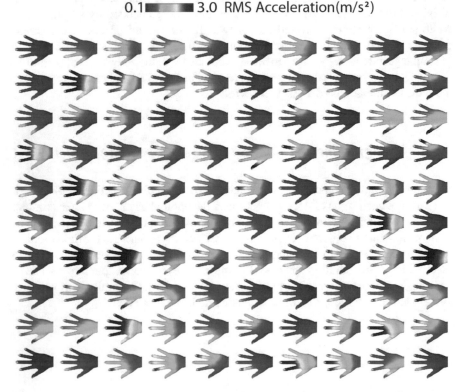

Fig. 4.7 A random sampling of one hundred tactile stimuli drawn from the dataset illustrates its diversity (time averages shown)

theoretically sound, involving a convex optimization. All classifiers used a radial basis function SVM kernel (width 5.0, selected using an independent validation set). We evaluated classification performance using a standard (tenfold) cross-validation method, with a 90% training and 10% testing data split.

In order to assess the between-subjects generalizability of these inferences, we performed a cross-subject validation, in which we trained a classifier on data from three participants and tested it on data from the fourth, and averaged the results across each left-out participant.

The information loss (cost function) of the model encoding the tactile signals was compared across different encoding ranks, using the RMS residual (NMF of spatial stimuli) and the KL divergence (CNMF of spatiotemporal stimuli).

The encoding efficiency was also evaluated by measuring the sparseness of the activation weights h. The model included a non-negativity constraint that encouraged sparseness. Sensory processing research has often connected neural representations with sparse codes [11, 24, 60, 62]. The Hoyer sparseness measure [21], a normalized ratio of ℓ^1 and ℓ^2 norms, is often preferred, based on criteria discussed in the literature [22]:

$$\text{Sparseness}(h) = \frac{\sqrt{n} - \left(\sum |h_i|\right)/\sqrt{\sum h_i^2}}{\sqrt{n} - 1}, \qquad (4.3)$$

where n is the number of elements in h, and i is the index of elements in h.

4.5.1 Classifying Manual Interactions from Tactile Codes

The encoding residual decreased with the number of bases. Five bases were sufficient to maximize the accuracy (80%) with which stimuli from one subject could be classified using only data from the other participants (Fig. 4.8). These five bases were highly conserved between subjects and were associated with individual digits. The activations of the bases exhibited a high degree of sparsity preserved across many trials (Table 4.1). Those that were associated with multiple finger contact were less sparse and more diverse than those involving just one finger. Information independence among the basis activations decreased with the number of digits engaged.

The encoding was not selected in order to optimize classification accuracy. Nonetheless, the classification accuracy increased with the number of bases, and was greater than 90% if at least seven bases were used, or greater than 95% if 12 were used. A high classification accuracy is not necessarily expected to be achieved when the number of input dimensions exceeds the number of classes. For example, there are binary prediction tasks, such as cancer prognosis prediction from images, where the consideration of multitudes of features is required to achieve moderately accurate classification [6].

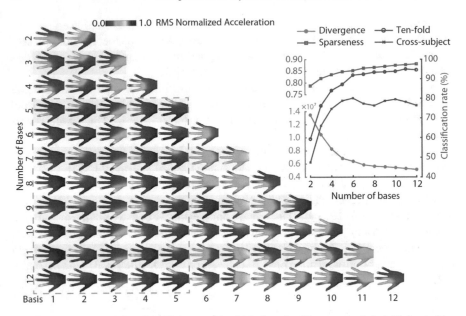

Fig. 4.8 Left: Maximizing the efficiency with which the stimuli were encoded yielded primitive bases, $w_i(x, t)$ (time averages shown). Each row corresponds to an encoding of fixed rank, M, from M = 2–12 bases, arranged in order of increasing activation. Individuated digit representations were highly conserved. Higher rank encodings included additional diffuse patterns. Top right: The encoding residual decreased with the rank, while classification improved

Table 4.1 Sparseness of the code and redundancy between basis activations for tactile stimuli elicited by all gestures. Results are displayed for an encoding with eight bases, and are qualitatively representative of results obtained with different numbers of bases

Gesture	Shannon entropy, H_J	Hoyer sparseness	$H_J - H_S$
Tap V	1.6	0.89	7.4
Tap IV	1.7	0.90	7.9
Tap I	1.8	0.90	8.5
Tap III	1.8	0.87	9.2
Tap II	1.9	0.87	9.6
Tap II, III	2.1	0.86	11.2
Grip C1 I, II	2.1	0.83	10.6
Grip C2 I, II	2.2	0.82	11.1
Slide II	2.2	0.82	11.5
Indirect tap I, II	2.3	0.79	11.8
Tap II, III, IV, V	2.5	0.85	14.0
Tap all	2.7	0.84	15.0
Grip ball All	3.0	0.78	16.8

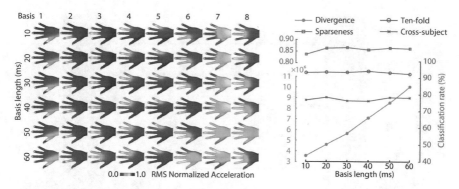

Fig. 4.9 Spatiotemporal bases obtained as the maximum duration is varied. Left: We encoded the stimuli with spatiotemporal bases of durations of 10–60 ms. The characteristic motifs were preserved in all cases, with the greatest changes in the most diffuse bases, which were least active in encoding the dataset (Fig. 4.8). Right: Classification accuracy, and cross-participant classification accuracy, varied little with basis duration. The linear increase in divergence with duration reflects the larger number of values that contributed to the measure

Fig. 4.10 Efficient codes of tactile waves yield sparse, additive descriptions of contact gestures. Optimal basis set for encoding time-averaged tactile waves with ℓ^0-sparsity

Similar patterns emerged when the encoding rank M, or number of bases, was adjusted (Fig. 4.8) or when optimizing with different initial conditions, data subsets, or optimization objectives (Figs. 4.9, 4.10, 4.11 and 4.12).

Information of Manual Interactions

In addition to sparseness (Eq. 4.3), the encoding efficiency of different hand interactions was also compared based on information entropy. We assessed the diversity of the encoding by computing the empirical Shannon information entropy of the activation values across the entire dataset, as follows. Discretize the activation values $h_i(t)$ and let p_k be the probability that a randomly drawn activation from any stimulus, channel, or time has a value lying in histogram bin k. The joint entropy of activations in all channels was

$$H_J = H(\{h_1, h_2, \ldots, h_M\}) = -\sum_{k=1}^{K} p_k \log_2 p_k .$$

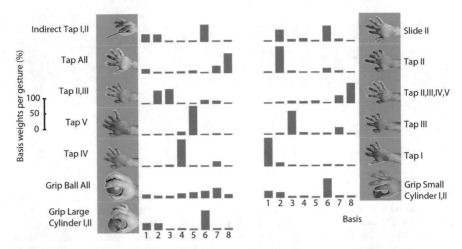

Fig. 4.11 Histograms of proportion of each basis contribution for each contact gesture, for learned bases (Fig. 4.10) with ℓ^0 coefficient sparsity constraints, which allowed three activated bases for each trial of the time-averaged data

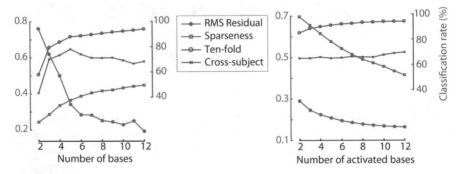

Fig. 4.12 Determining number of bases and activation constraints from gesture classifications. Left: The rate of correct classification of the contact gestures increased as the number of bases grew. Cross-individual classification rate peaked at five bases. Right: As the subset of activated bases grew from 2 to 12, the rate of classification increased by less than 15%

The distribution, p_k, was computed from all values of $h_i(t)$ in the entire dataset, for all i. The entropy H_J was maximized when all weight values were equally likely and decreased as the sparseness of the code increased. This measure revealed differences between encoded stimuli produced by different actions (Table 4.1). The joint activation entropy H_J was highest for gestures involving multi-finger contact, and lowest for contacts of single fingers, suggesting that the model was most efficient at encoding gestures involving individual digits. For each basis, we computed the entropies, $H(h_i)$,

$$H(h_i) = -\sum_k p(h_{i,k}) \log_2 p(h_{i,k}), \quad i = 1, \ldots, M,$$

where $h_{i,k}$ are the discrete values of the activation coefficients, h_i, from the histogram distribution, $p(h_{i,k})$, computed using the entire dataset. These entropies were then summed,

$$H_S = \sum_{i=1}^{M} H(h_i),$$

giving a result that satisfied, $H_J \leq H_S$. Equality was reached if the activations for different bases were statistically independent, thus, $H_S - H_J$ measured the degree of dependence of the activations among the bases (Table 4.1). When this quantity was minimized, the bases represented maximally independent components of the data.

4.5.2 Evaluating the Sparseness of the Encoding

An encoding that is excessively sparse may lead to loss of information. We thus enforced sparsity in the encoding and evaluated the impact at different sparseness levels.

Spatial Tactile Stimuli

We first investigated the effects of sparseness on the encoding of time-averaged data. To encourage sparseness, we restricted the number, M_H, of combined bases that could be activated on each trial. To this end, a sparsity-promoting ℓ^0 constraint, $C_0(\mathbf{H})$, was added to the objective function to be optimized. It had the form,

$$C_0(\mathbf{H}) = \lambda \left(|\mathbf{H}|_0 - M_H \right), \quad |\mathbf{H}|_0 = \text{Cardinality}\{H_{ij} \mid H_{ij} > 0\},$$

where λ is a regularization weight and H_{ij} are entries of the activation weight matrix $\mathbf{H}$ of the encoded dataset. The spatial patterns associated with those bases were similar to those obtained without the sparsity term (Fig. 4.10, with $M_H = 3$), and the average activation patterns across all trials indicated that tactile stimuli within each manual gesture category were consistently encoded with the same bases (Fig. 4.11). The exception was the multi-finger "Grip Ball" action, which elicited stimuli that were encoded with different combinations of the primitive bases.

We varied the rank of the decomposition, or total number of bases, M, the ℓ^0 sparsity values M_H, and assessed information content in the encoded signals via gesture classification using both tenfold cross-validation and cross-participant evaluation, where the data from one participant was used to evaluate the classifier trained using data from other participants (Fig. 4.12).

As the rank, M, increased from 3 to 12, mean classification accuracy rose by less than 15%, while mean cross-participant classification accuracy remained unchanged. As the ℓ^0-sparseness increased, leading to a growth in the number of activated bases from 3 to 12 per trial, the classification accuracy improved by only 5%, and cross-participant classification accuracy improved by only 3%. When the number of activated bases was reduced from three to two, cross-participant classification remained unchanged. These results show that the data was sparsely encoded using as few as two bases and that the captured information generalized across participants. Increasing the sparsity of the encoding of time-average, spatial stimuli had only a modest effect on encoding quality, as measured by classification accuracy within and between participants.

Spatiotemporal Tactile Stimuli

As in the case of the time-averaged NMF encoding model described earlier, we sought to investigate the tradeoff between coding sparsity and accuracy. To this end, inspired by prior studies of speech coding [49], the CNMF algorithm was modified to require that it achieves a specified value of the Hoyer sparseness measure (Eq. 4.3). We used an objective function of the form,

$$D = \sum_{x,t} \left| a(\mathbf{x}, t) - \sum_{i=1}^{M} \sum_{\tau=0}^{T} h_i(t - \tau) \, w_i(x, \tau) \right|^2 \quad \text{such that} \quad \text{Sparseness}\{h_i(t)\} = S_\mathrm{H},$$

where $S_\mathrm{H} \in [0, 1]$ is a predetermined Hoyer sparseness. Here, the approximation quality was measured via the squared residual error between the stimulus, $v(\mathbf{x}, t)$, and the convolutional product of h_i and w_i rather than the KL divergence. The Hoyer sparseness constraint, S_H, included all basis activations, and was enforced for each trial.

The bases remained qualitatively similar as the imposed sparseness varied (Fig. 4.13), until the sparseness value, S_H, approached one, when the patterns became progressively diffuse. By doing so, they could approximate the effect of activations involving larger hand regions, which could also be achieved through the activation of more bases, at the expense of sparseness.

When we evaluated the encodings using the manual action classification task described above, the rate of correct classification decreased for sparseness values greater than 0.8 (Fig. 4.8), indicating that information was lost if the encoding was required to be very sparse.

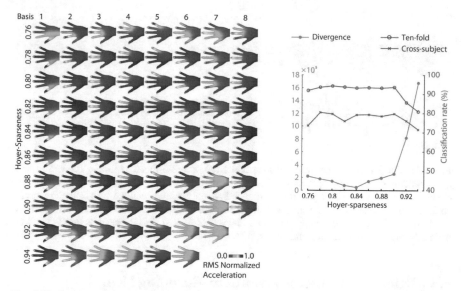

Fig. 4.13 Spatiotemporal bases obtained under constraints of varying sparseness. Left: RMS time averages shown. For modest levels of sparseness, the spatial patterns were similar as sparseness varied, while for sparseness values greater than about 0.86, several bases lost definition, becoming more diffuse, while classification accuracy also decreased (Fig. 4.8). Right: The KL divergence increased with sparseness, with little effect on classification accuracy until sparseness values exceeded 0.9

4.5.3 Encoding Robustness

It could be hypothesized that the structure of our dataset favored such an encoding, in which several spatiotemporal basis patterns were associated with individual digits. For example, 45% of the 4,600 analyzed tactile stimuli were elicited by gestures that produced contact at only one digit. To investigate this possibility, we applied the same analysis to a subset of the data that excluded tactile stimuli produced by single-digit gestures. We repeated the same analysis using an additional subset that only included stimuli produced via gestures involving contact with all five digits. In each case, the results (Fig. 4.14) were highly similar to those we obtained from encoding the entire dataset (Fig. 4.8), including distinct basis patterns that were primarily localized in single digits.

0.0 ▬▬ 1.0 RMS Normalized Acceleration

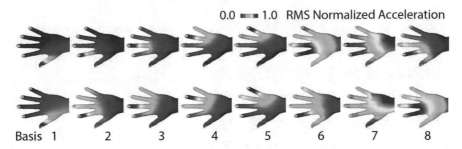

Basis 1   2   3   4   5   6   7   8

Fig. 4.14 Spatiotemporal encoding of stimuli elicited by multi-finger gestures. Top row: Tactile basis patterns obtained by optimally encoding a subset of stimuli that excluded all single-digit contacts. Bottom row: Basis patterns from encoding only stimuli elicited by gestures involving contact with all fingers (time averages shown). In both cases, the results were highly similar to those we obtained from encoding the entire dataset, including distinct basis patterns that were primarily localized in individual digits

4.6   Relevance to Tactile Sensation: Encoding Neural Population Responses via Spiking Simulations

The observed encoding efficiency of the mechanical signals was a consequence of spatiotemporal integration supplied by the tactile basis patterns. Prevailing physiological models leave little doubt that the spatial and temporal properties of touch-elicited mechanical signals are reflected in the volleys of afferent activations during natural hand interactions. However, no method is known to allow simultaneous capture of neural signals from populations of peripheral afferents in the behaving hand. We instead employed a biologically justified neuron spiking simulation software (TouchSim [51]). This software predicted the firing patterns of a population of 773 vibration-sensitive afferents distributed throughout a simulated hand in response to the touch-elicited vibrations of the skin that we captured in vivo.

We computed skin displacements from the skin acceleration data. The skin displacement signals were used to drive the TouchSim model. For each of the 4,600 trials of the entire database, the model produced and output spike trains for each of the 773 simulated PC afferents (PC afferents form a class of afferent fibers terminating in Pacinian corpuscles thought to play a major role in the encoding of skin vibrations [9, 27, 58]). We also computed mean firing rates for each. The output of the simulation was thus the spike train data and mean firing rate of the afferents for each trial. The firing patterns of some representative trials are shown in Fig. 4.15. Mean firing rates and spike train data for gestures tapping either digit II or III are shown in Fig. 4.16, where the selected afferents (colored) were ordered based on their distance from the location of the contact stimulus. The spiking data exhibits an oscillating pattern whose phase increases with distance from the fingertip. We attributed to this the propagation dynamics of elastic waves in the skin, consistent with theory (Sect. 4.8) and mechanical measurements.

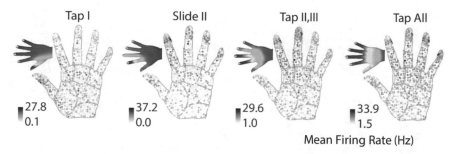

Tap I   Slide II   Tap II,III   Tap All

■27.8   ■37.2   ■29.6   ■33.9
 0.1   0.0   1.0   1.5

Mean Firing Rate (Hz)

Fig. 4.15 Spiking rate of populations of simulated mechanosensory afferents. Responses of FA-II (PC) afferents. The firing rates were averaged over the 800 ms duration for each trial

Fig. 4.16 Responses of simulated FA-II afferents along the bone axis of the tapping digit (duration shown: 28 ms). The firing patterns of selected afferents are shown in the raster plot, corresponding to the colored region

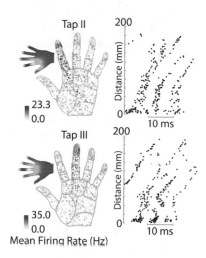

Tap II   200

Distance (mm)

■23.3
 0.0   0
                          10 ms

Tap III   200

Distance (mm)

■35.0
 0.0   0
                          10 ms

Mean Firing Rate (Hz)

We optimized the encoding of the predicted neural responses with the similar method that was employed for mechanical signals: analyzing the mean firing rate data for all 773 PCs using a NMF procedure. Informed by our analysis of the acceleration data, we performed the NMF analysis of the simulated neural data for 2–12 bases, yielding 11 different encodings of increasing dimensionality. The 8-basis solution is shown in Fig. 4.17. Each basis describes a distribution of mean firing rates used in the encoding. The bases bore a striking resemblance to those that we obtained by analyzing the acceleration data (Fig. 4.8). We evaluated the quality of the encodings using a classification task, residual measure, sparseness measure, and a cross-subject classification task. The classification and evaluation procedures were exactly the same as those used for the accelerometer data and were not optimized for this data. Nonetheless, classification rates reached 90% with 8 bases (Fig. 4.17).

The spiking bases exhibited similar patterns of spatial integration to the bases we obtained using the mechanical data, including individuation of digit representations and denser activation of the fingertips. The results of the classification tasks were qualitatively similar to those that we obtained from the mechanical data, despite the

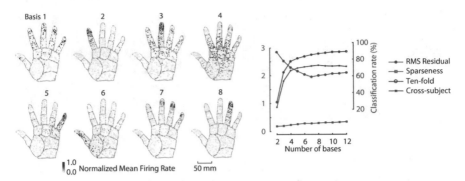

Fig. 4.17 Left: Optimizing the encoding of spiking data from 773 simulated afferents yielded bases (M = 8 shown) that integrated activity in neural populations throughout the hand. They reflected the individuation of digits and denser activation of the fingertips, similar to the results obtained with the mechanical data. Right: Results based on the neural simulations. As the number of spiking bases increased, the encoding residual decreased, and classification also improved

higher dimensionality of the input data. This suggests that the encoding revealed organizational principles that would be preserved by neurotransduction, and that went beyond the mere properties of skin vibrations.

4.7 Discussion

The size and the diversity of our corpus of data were limited by experimental constraints, as in many other studies utilizing corpora of motor or sensory data [34, 53, 55]. Although we selected the gestures based on the most reasonable assumptions available, a larger dataset could be captured during spontaneous manual activities outside of the lab, or specified based on analyses of conditions in which our species evolved.

A useful comparison can be drawn to research on hand movements and grasping. Research in this area has shown how a relatively small number of coordination patterns ("synergies") can explain most of the variability in hand movement data. Similar coordination patterns have been observed in studies based on different laboratory datasets, or on spontaneous manual activities outside the lab, with some task dependency [53]. The dimensionality of the analysis presented here is much larger than that typically arises in hand kinematics studies. Nonetheless, analogous considerations may apply to our findings. Indeed, our analyses of subsets of the mechanical data (Fig. 4.14) yielded basis patterns that were very similar to those we obtained from the combined dataset.

The tactile basis patterns were also invariably organized along a gradient from higher to lower finger individuation. This is opposite to the trend that is observed in grasping studies, and may evince an important difference in organizational principles between the tactile and motor systems. The larger degree of individuation that our findings suggest appears to be a consequence of the physics of vibration transmission in the skin, which causes propagating vibrations to attenuate with increasing distance from their source. In contrast, hand movement studies reveal a higher degree of multi-digit coordination, which is facilitated by the biomechanics of the limb.

While the behavioral relevance of propagating vibrations in the limb is not fully understood, prior research shows how these vibrations can mediate tactile perception [35, 41]. Further research is needed in order to clarify the relevance of the predictions from efficient tactile encoding to hand function and somatosensory processing.

4.8 Conclusion

Our findings suggest that the biomechanics of the hand can facilitate tactile perception by affecting the preneuronal compression of tactile information in the whole hand. This compression was produced by a compact lexicon of primitive tactile wave patterns. Spatiotemporal integration supplied by these basis patterns optimally encoded the tactile stimuli. Recent studies of neural correlates of somatosensory processing reveal that at the earliest stages of cortical processing, individual neurons exhibit complex responses to tactile stimuli distributed throughout the extremities [15, 17, 36]. These studies, together with the new findings presented here, show how traditional depictions of receptive fields do not reflect the extent of early somatosensory integration [50], including effects of mechanical transmission in the body.

In summary, this chapter shows that skin vibrations elicited by hand interactions produce an efficient encoding of tactile information about the interactions. We obtained a compact lexicon of primitive wave patterns from the computation of an optimal encoding of thousands of naturally occurring tactile stimuli. The encoding sparsely represented the entire dataset, enabling touch interactions to be classified with an accuracy exceeding 95%. The primitive tactile patterns reflected the interplay of hand anatomy with wave physics. Similar patterns emerged when we applied efficient encoding criteria to spiking data from populations of simulated tactile afferents. This finding suggests that the biomechanics of the hand enables efficient perceptual processing by effecting a preneuronal compression of tactile information.

Appendix: Mechanical Waves in the Tactile Regime

Time-varying loads applied to the skin elicit mechanical waves that are transmitted via viscoelastic coupling of skin and other tissues. Waves with frequencies lower than few hundred hertz and amplitude less than 0.1 mm are shown to propagate primarily along the skin surface [45]. The part of shear waves propagating in depth through the tissue attenuate rapidly is restricted to the stimulation area [40]. A physical model was proposed to describe the propagation of the low-frequency ($\leq$1 kHz) surface waves, assuming the skin to be a semi-infinite, locally isotropic, elastic continuum [30]. In this book, we captured surface wave propagation on the skin using wearable accelerometers and vibrometers. We focus on small skin displacement elicited by hand interactions. The perceptual relevant range of the skin vibration is below 1 kHz, as described in the previous section. Thus, we consider the skin as an idealized elastic medium and adopt the same model for surface wave propagation on the skin.

Consider a time-varying strain vector $\boldsymbol{\xi}(\mathbf{x}, t)$ denoting the localized displacement at location $\mathbf{x}$ on the skin, at time t. The governing equation for the idealized medium is

$$(\lambda + \mu)\nabla(\nabla \cdot \boldsymbol{\xi}(\mathbf{x}, t)) + \mu\nabla^2\boldsymbol{\xi}(\mathbf{x}, t) = \rho\frac{\partial^2\boldsymbol{\xi}(\mathbf{x}, t)}{\partial t^2}, \tag{4.4}$$

where ∇ is the Laplace operator, ρ is the density of the medium, λ and μ are the Lame constants of the medium. μ is the shear modulus under the context of elasticity [52]. By virtue of the isotropic property of this idealized model, the Lame's first parameter $\lambda = K - \frac{2}{3}\mu$, where K is the bulk modulus [20]. Therefore, the governing equation can be rewritten as

$$\left(K + \frac{\mu}{3}\right)\nabla(\nabla \cdot \boldsymbol{\xi}(\mathbf{x}, t)) + \mu\nabla^2\boldsymbol{\xi}(\mathbf{x}, t) = \rho\frac{\partial^2\boldsymbol{\xi}(\mathbf{x}, t)}{\partial t^2}. \tag{4.5}$$

ρ, μ, K are scalar constants due to the homogeneous property of the medium.

One solution to the above governing equation is

$$\boldsymbol{\xi}_f(\mathbf{x}, t) = e^{j(\mathbf{k}\cdot\mathbf{x} - \omega t)} = e^{j(\mathbf{k}\cdot\mathbf{x} - 2\pi f t)}, \tag{4.6}$$

where $\mathbf{k}$ is the wave vector indicating the propagation direction of the wave, and $\omega = 2\pi f$ is the angular frequency [61]. This solution describes a plane wave of a single frequency f (monochromatic). Thus, the derivatives can be replaced and rewritten as

$$\left(K + \frac{\mu}{3}\right)\mathbf{k}(\mathbf{k} \cdot \boldsymbol{\xi}) + \mu k^2\boldsymbol{\xi} = \rho\omega^2\boldsymbol{\xi}, \tag{4.7}$$

where $k = |\mathbf{k}|$ is the wave number, representing the three-dimensional direction of the wave, with unit direction $\hat{\mathbf{k}}$. The strain vector $\boldsymbol{\xi}$ indicates the displacement of the medium along three orthogonal directions. The oscillation of $\boldsymbol{\xi}$ is parametrized by the wave vector $\mathbf{k}$ and angular frequency of the wave ω. Let $\hat{\mathbf{k}}$ be the unit-length vector denoting the direction of $\mathbf{k}$. By projecting $\boldsymbol{\xi}$ on different directions, elastic waves $\boldsymbol{\xi}$ can be decomposed to two components that are orthogonal to each other: longitudinal one ξ_L and transverse one $\boldsymbol{\xi}_T$ ($\boldsymbol{\xi} = \boldsymbol{\xi}_L + \boldsymbol{\xi}_T$). The longitudinal component describes the body expansion of the medium, with displacement in parallel to the direction of wave propagation, namely $\xi_L \times \hat{\mathbf{k}} = 0$. In contrast, the transverse component represents the shear waves with displacement point in directions orthogonal to $\mathbf{k}$, namely $\boldsymbol{\xi}_T \cdot \hat{\mathbf{k}} = 0$. Thus, $\boldsymbol{\xi}_T$ is a two-dimensional vector.

Based on the decomposition of the waves, we have governing equations for the longitudinal waves: $\rho\omega^2 = (K + \frac{4}{3}\mu)k^2$ and that for the transversal shear waves: $\rho\omega^2 = \mu k^2$, which determine the propagation speed of the waves. For a monochromatic wave propagating in the medium, its phase velocity $v = \frac{\omega}{k}\hat{\mathbf{k}}$, and its phase speed $c = |v| = \frac{\omega}{k}$. The phase speed of longitudinal waves, $c_L = \sqrt{(K + \frac{4}{3}\mu)/\rho}$, does not depend on the wave number k nor frequency f of the waves. The phase speed of (transversal) shear waves $c_T = \sqrt{\mu/\rho}$. $c_T < c_L$ since the bulk modulus $K > 0$. Prior studies found that shear waves travel at low speeds, $1 < c_T < 30$ m/s in soft tissues, while longitudinal waves travel at a much fast speed, $c_L \approx 1730$ m/s in the skin and $c_L \approx 1575$ m/s in the muscle [2, 40]. Surface waves propagating on the skin have speeds similar to those of shear waves.

References

1. Attneave, F.: Some informational aspects of visual perception. Psychol. Rev. **61**(3), 183 (1954)
2. Azhari, H.: Basics of biomedical ultrasound for engineers. Wiley (2010)
3. Bagdasarian, K., Szwed, M., Knutsen, P.M., Deutsch, D., Derdikman, D., Pietr, M., Simony, E., Ahissar, E.: Pre-neuronal morphological processing of object location by individual whiskers. Nat. Neurosci. **16**(5), 622 (2013)
4. Barlow, H.B., et al.: Possible principles underlying the transformation of sensory messages. Sens. Commun. **1**, 217–234 (1961)
5. Bashivan, P., Rish, I., Yeasin, M., Codella, N.: Learning representations from EEG with deep recurrent-convolutional neural networks (2015). arXiv:1511.06448
6. Beck, A.H., Sangoi, A.R., Leung, S., Marinelli, R.J., Nielsen, T.O., Van De Vijver, M.J., West, R.B., Van De Rijn, M., Koller, D.: Systematic analysis of breast cancer morphology uncovers stromal features associated with survival. Sci. Transl. Med. **3**(108), 108ra113–108ra113 (2011)
7. Bell, A.J., Sejnowski, T.J.: The "independent components" of natural scenes are edge filters. Vis. Res. **37**(23), 3327–3338 (1997)
8. Bengtsson, F., Brasselet, R., Johansson, R.S., Arleo, A., Jörntell, H.: Integration of sensory quanta in cuneate nucleus neurons in vivo. PloS One **8**(2) (2013)
9. Cauna, N., Mannan, G.: The structure of human digital pacinian corpuscles (corpuscula lamellosa) and its functional significance. J. Anat. **92**(Pt 1), 1 (1958)
10. Chatterjee, M., Zwislocki, J.J.: Cochlear mechanisms of frequency and intensity coding. i. the place code for pitch. Hear. Res. **111**(1-2), 65–75 (1997)
11. Crochet, S., Poulet, J.F., Kremer, Y., Petersen, C.C.: Synaptic mechanisms underlying sparse coding of active touch. Neuron **69**(6), 1160–1175 (2011)
12. Dan, Y., Atick, J.J., Reid, R.C.: Efficient coding of natural scenes in the lateral geniculate nucleus: experimental test of a computational theory. J. Neurosci. **16**(10), 3351–3362 (1996)
13. De Boer, E., Viergever, M.: Wave propagation and dispersion in the cochlea. Hear. Res. **13**(2), 101–112 (1984)
14. Delhaye, B., Hayward, V., Lefèvre, P., Thonnard, J.L.: Texture-induced vibrations in the forearm during tactile exploration. Front. Behav. Neurosci. **6**, 37 (2012)
15. Enander, J.M., Jörntell, H.: Somatosensory cortical neurons decode tactile input patterns and location from both dominant and non-dominant digits. Cell Rep. **26**(13), 3551–3560 (2019)
16. Field, D.J.: What is the goal of sensory coding? Neural Comput. **6**(4), 559–601 (1994)
17. Foffani, G., Chapin, J.K., Moxon, K.A.: Computational role of large receptive fields in the primary somatosensory cortex. J. Neurophysiol. **100**(1), 268–280 (2008)
18. Gallardo, A.P., Epp, B., Dau, T.: Can place-specific cochlear dispersion be represented by auditory steady-state responses? Hear. Res. **335**, 76–82 (2016)

19. Gardner, E.P., Costanzo, R.M.: Spatial integration of multiple-point stimuli in primary somatosensory cortical receptive fields of alert monkeys. J. Neurophysiol. **43**(2), 420–443 (1980)
20. Greenleaf, J.F., Fatemi, M., Insana, M.: Selected methods for imaging elastic properties of biological tissues. Annu. Rev. Biomed. Eng. **5**(1), 57–78 (2003)
21. Hoyer, P.O.: Non-negative matrix factorization with sparseness constraints. J. Mach. Learn. Res. **5**(Nov), 1457–1469 (2004)
22. Hurley, N., Rickard, S.: Comparing measures of sparsity. IEEE Trans. Inf. Theory **55**(10), 4723–4741 (2009)
23. Iwamura, Y., Tanaka, M., Sakamoto, M., Hikosaka, O.: Rostrocaudal gradients in the neuronal receptive field complexity in the finger region of the alert monkey's postcentral gyrus. Exp. Brain Res. **92**(3), 360–368 (1993)
24. Jadhav, S.P., Wolfe, J., Feldman, D.E.: Sparse temporal coding of elementary tactile features during active whisker sensation. Nat. Neurosci. **12**(6), 792 (2009)
25. Johansson, R.S., Birznieks, I.: First spikes in ensembles of human tactile afferents code complex spatial fingertip events. Nat. Neurosci. **7**(2), 170–177 (2004)
26. Johansson, R.S., Flanagan, J.R.: Coding and use of tactile signals from the fingertips in object manipulation tasks. Nat. Rev. Neurosci. **10**(5), 345–359 (2009)
27. Johansson, R.S., Landstro, U., Lundstro, R., et al.: Responses of mechanoreceptive afferent units in the glabrous skin of the human hand to sinusoidal skin displacements. Brain Res. **244**(1), 17–25 (1982)
28. Kim, D.H., Baddar, W.J., Jang, J., Ro, Y.M.: Multi-objective based spatio-temporal feature representation learning robust to expression intensity variations for facial expression recognition. IEEE Trans. Affect. Comput. **10**(2), 223–236 (2017)
29. Kim, J., Truong, K.P., Englebienne, G., Evers, V.: Learning spectro-temporal features with 3D CNNs for speech emotion recognition. In: 2017 Seventh International Conference on Affective Computing and Intelligent Interaction (ACII), pp. 383–388. IEEE (2017)
30. Kirkpatrick, S.J., Duncan, D.D., Fang, L.: Low-frequency surface wave propagation and the viscoelastic behavior of porcine skin. J. Biomed. Opt. **9**(6), 1311–1320 (2004)
31. Lee, D.D., Seung, H.S.: Learning the parts of objects by non-negative matrix factorization. Nature **401**(6755), 788–791 (1999)
32. Lee, D.D., Seung, H.S.: Algorithms for non-negative matrix factorization. In: Advances in Neural Information Processing Systems, pp. 556–562 (2001)
33. Lewicki, M.S.: Efficient coding of natural sounds. Nat. Neurosci. **5**(4), 356–363 (2002)
34. Lewicki, M.S., Sejnowski, T.J.: Coding time-varying signals using sparse, shift-invariant representations. In: Advances in Neural Information Processing Systems, pp. 730–736 (1999)
35. Libouton, X., Barbier, O., Berger, Y., Plaghki, L., Thonnard, J.L.: Tactile roughness discrimination of the finger pad relies primarily on vibration sensitive afferents not necessarily located in the hand. Behav. Brain Res. **229**(1), 273–279 (2012)
36. Lipton, M.L., Liszewski, M.C., O'Connell, M.N., Mills, A., Smiley, J.F., Branch, C.A., Isler, J.R., Schroeder, C.E.: Interactions within the hand representation in primary somatosensory cortex of primates. J. Neurosci. **30**(47), 15895–15903 (2010)
37. Liu, J.X., Wang, D., Gao, Y.L., Zheng, C.H., Xu, Y., Yu, J.: Regularized non-negative matrix factorization for identifying differentially expressed genes and clustering samples: a survey. IEEE/ACM Trans. Comput. Biol. Bioinf. **15**(3), 974–987 (2017)
38. Manfredi, L.R., Baker, A.T., Elias, D.O., Dammann III, J.F., Zielinski, M.C., Polashock, V.S., Bensmaia, S.J.: The effect of surface wave propagation on neural responses to vibration in primate glabrous skin. PloS One **7**(2) (2012)
39. Manfredi, L.R., Saal, H.P., Brown, K.J., Zielinski, M.C., Dammann, J.F., III., Polashock, V.S., Bensmaia, S.J.: Natural scenes in tactile texture. J. Neurophysiol. **111**(9), 1792–1802 (2014)
40. Moore, T.J.: A survey of the mechanical characteristics of skin and tissue in response to vibratory stimulation. IEEE Trans. Man-Mach. Syst. **11**(1), 79–84 (1970)

41. Morley, J., Hawken, M., Burge, P.: Vibratory detection thresholds following a digital nerve lesion. Exp. Brain Res. **72**(1), 215–218 (1988)
42. Neimark, M.A., Andermann, M.L., Hopfield, J.J., Moore, C.I.: Vibrissa resonance as a transduction mechanism for tactile encoding. J. Neurosci. **23**(16), 6499–6509 (2003)
43. Nobili, R., Mammano, F., Ashmore, J.: How well do we understand the cochlea? Trends Neurosci. **21**(4), 159–167 (1998)
44. Olshausen, B.A., Field, D.J.: Emergence of simple-cell receptive field properties by learning a sparse code for natural images. Nature **381**(6583), 607–609 (1996)
45. Potts, R.O., Chrisman, D.A., Jr., Buras, E.M., Jr.: The dynamic mechanical properties of human skin in vivo. J. Biomech. **16**(6), 365–372 (1983)
46. Prsa, M., Morandell, K., Cuenu, G., Huber, D.: Feature-selective encoding of substrate vibrations in the forelimb somatosensory cortex. Nature **567**(7748), 384–388 (2019)
47. Pubols, B.H.: Effect of mechanical stimulus spread across glabrous skin of raccoon and squirrel monkey hand on tactile primary afferent fiber discharge. Somat. Res. **4**(4), 273–308 (1987)
48. Qiu, Z., Yao, T., Mei, T.: Learning spatio-temporal representation with pseudo-3D residual networks. In: proceedings of the IEEE International Conference on Computer Vision, pp. 5533–5541 (2017)
49. Ramanarayanan, V., Goldstein, L., Narayanan, S.S.: Spatio-temporal articulatory movement primitives during speech production: Extraction, interpretation, and validation. J. Acoust. Soc. Am. **134**(2), 1378–1394 (2013)
50. Reed, J.L., Pouget, P., Qi, H.X., Zhou, Z., Bernard, M.R., Burish, M.J., Haitas, J., Bonds, A., Kaas, J.H.: Widespread spatial integration in primary somatosensory cortex. Proc. Natl. Acad. Sci. **105**(29), 10233–10237 (2008)
51. Saal, H.P., Delhaye, B.P., Rayhaun, B.C., Bensmaia, S.J.: Simulating tactile signals from the whole hand with millisecond precision. Proc. Natl. Acad. Sci. **114**(28), E5693–E5702 (2017)
52. Salençon, J.: Handbook of continuum mechanics: general concepts thermoelasticity. Springer Science & Business Media (2012)
53. Santello, M., Baud-Bovy, G., Jörntell, H.: Neural bases of hand synergies. Front. Comput. Neurosci. **7**, 23 (2013)
54. Shao, Y., Hayward, V., Visell, Y.: Spatial patterns of cutaneous vibration during whole-hand haptic interactions. Proc. Natl. Acad. Sci. **113**(15), 4188–4193 (2016)
55. Simoncelli, E.P., Olshausen, B.A.: Natural image statistics and neural representation. Annu. Rev. Neurosci. **24**(1), 1193–1216 (2001)
56. Smaragdis, P.: Non-negative matrix factor deconvolution; extraction of multiple sound sources from monophonic inputs. In: International Conference on Independent Component Analysis and Signal Separation, pp. 494–499. Springer (2004)
57. Sotiras, A., Resnick, S.M., Davatzikos, C.: Finding imaging patterns of structural covariance via non-negative matrix factorization. Neuroimage **108**, 1–16 (2015)
58. Stark, B., Carlstedt, T., Hallin, R., Risling, M.: Distribution of human pacinian corpuscles in the hand: a cadaver study. J. Hand Surg. **23**(3), 370–372 (1998)
59. Sur, M.: Receptive fields of neurons in areas 3b and 1 of somatosensory cortex in monkeys. Brain Res. **198**(2), 465–471 (1980)
60. Terashima, H., Hosoya, H., Tani, T., Ichinohe, N., Okada, M.: Sparse coding of harmonic vocalization in monkey auditory cortex. Neurocomputing **103**, 14–21 (2013)
61. Thorne, K.S., Blandford, R.D.: Modern Classical Physics: Optics, Fluids, Plasmas, Relativity, and Statistical Physics. Princeton University Press, Elasticity (2017)
62. Vinje, W.E., Gallant, J.L.: Sparse coding and decorrelation in primary visual cortex during natural vision. Science **287**(5456), 1273–1276 (2000)
63. Von Békésy, G., Wever, E.G.: Experiments in hearing, vol. 8. McGraw-Hill New York (1960)

64. Vu, T.T., Bigot, B., Chng, E.S.: Speech enhancement using beamforming and non negative matrix factorization for robust speech recognition in the chime-3 challenge. In: 2015 IEEE Workshop on Automatic Speech Recognition and Understanding (ASRU), pp. 423–429. IEEE (2015)
65. Weber, A.I., Saal, H.P., Lieber, J.D., Cheng, J.W., Manfredi, L.R., Dammann, J.F., Bensmaia, S.J.: Spatial and temporal codes mediate the tactile perception of natural textures. Proc. Natl. Acad. Sci. **110**(42), 17107–17112 (2013)
66. Yang, Y., Aminoff, E., Tarr, M., Robert, K.E.: A state-space model of cross-region dynamic connectivity in MEG/EEG. In: Advances in Neural Information Processing Systems, pp. 1234–1242 (2016)

Chapter 5
A Wearable Tactile Sensor Array for Large Area Remote Vibration Sensing in the Hand

Abstract In order to engineer haptic technologies for the hand, knowledge about what signals are felt during natural interactions is needed. The findings from Chaps. 3 and 4 demonstrate the utility of examining distributed tactile vibrations accompanying whole-hand haptic interactions. However, existing sensing devices cannot capture the full range of tactile information in the naturally behaving hand and are unable to match human abilities of perception and action. Thus, in this chapter presents the design of a new sensor apparatus, comprising a 126-channel wearable tactile sensor array, that is adapted to the anatomy of the hand, and that corresponds to the frequency sensitivity range of human tactile sensing. This device permits tactile sensing in vivo without kinematic constraints on hand movements. It provides new methods for collecting tactile data outside of constrained laboratory experiments, physiologically informed signal processing methods for reconstructing whole-hand tactile signals, and new methods for distributed tactile sensing that may have applications in robotics, upper-limb prosthetics, and other domains.

Disclaimer

Previously published as: Y. Shao, H. Hu, and Y. Visell, A Wearable Tactile Sensor Array for Large Area Remote Vibration Sensing in the Hand. *IEEE Sensors Journal*, Feb 2020, 20 (12), 6612–6623; DOI: 10.1109/JSEN.2020.2972521. Reproduced here by permission of IEEE.

5.1 Introduction

The human hand is endowed with many thousands of mechanosensory neurons that reach densities exceeding $10/mm^2$ of skin in the fingertip [16]. They enable us to perform remarkable perceptual feats, from easily discriminating extremely fine differences in surface texture and materials, to detecting nanometer-scale surface features [26]. Such perceptual tasks rely on the low-level transduction of mechanical signals spanning a wide frequency range (from approximately 0 to 1000 Hz) and dynamic range (from 10^{-3} to 10^{-8} m) into electrical spikes that are transmitted to the brain. This neural information gives rise to conscious experiences of touching objects and surfaces and enables the great variety of abilities of the hand, including object grasping, haptic perception, and manipulation [15].

The human hand is thus a remarkable sensory and prehensile instrument that provides a biological model for engineering systems for robotic manipulation and tactile sensing. There is growing interest in the engineering of tactile sensors that might be able to reproduce the remarkable feats of perception of biological skin, and that may emulate the large range of motor functions of the human hand. In this research, we present a new wearable tactile sensing array, with which we aim to provide quantitative information about the tactile signals that are captured by the human hand during natural interactions. This system also provides a new approach for skin-like artificial sensing that is inspired by the distributed vibration sensing capacities that are intrinsic to the human hand. It is based on a distributed array of sensors that, in our device, are elastically coupled through the soft tissues of the hand.

The multi-modal nature of touch sensing in the hand presents challenges that have made it difficult to fully understand the mechanisms of tactile sensing. It also provides inspiration for the design of biologically informed artificial tactile sensors that have not been fully exploited to date. There are more than a half dozen types of sensory receptors of touch in humans and other mammals [27], capturing different aspects and components of the mechanical signals supporting touch sensing and interaction [50]. Subpopulations of these receptors are more responsive to sustained mechanical stimulation (slowly adapting types, SA), and others to transient or vibratory signals (fast adapting, FA). Receptors can be further categorized as sensitive to stimulation near to a mechanical contact (types SA I and FA I), or to more remote mechanical signals (SA II or FA II). Essentially all perceptual and motor functions of the human hand are enabled by input from several of these tactile submodalities, yielding a variety of information about skin–object contact [17, 34].

While conventional accounts of biological touch sensing associate tactile perception with sensory resources near to the area of contact with an object (since these are the tissues that undergo the largest contact-induced deformations), recent research, including work in our own lab, reveals that mechanical contact with the skin elicits elastic waves that propagate throughout the hand in the tactile frequency range, from 0 to 800 Hz. These mechanical signals reach widespread vibration-sensitive receptors (including type FA II receptors, also called Pacinian corpuscles, PC), where they are transduced into neural signals that can encode contact forces, events, or surfaces with which they originate [5, 25, 38]. This remote transmission of touch-elicited mechanical signals in the hand is very efficient at frequency ranges (approximately 20–800 Hz) that are relevant to vibration perception [5, 38].

Human abilities of remote tactile sensing are especially associated with FA-II receptors (Pacinian corpuscles, PC). PC numbers in the many hundreds in each hand [18] are distributed throughout the limb, and are involved in tactile functions including texture discrimination, tool use, and the detection of object contact or slip.

However, it has been difficult to characterize the mechanical processes involved in human remote tactile sensing due to the complex array of tissues [20] and physical regimes involved.

Few electronic systems have been designed to capture the array of distributed touch-related mechanical signals in the whole hand during natural tactile exploration, grasping, and manipulation in everyday activities. As a result, we currently have a limited understanding of tactile signals that are felt during such interactions. Knowledge of this type could help to elucidate the scientific basis of human touch sensing, as mediated by continuum mechanics of the hand, including effects of sensory loss accompanying disease. It could also lead to advances in sensor engineering for robotics, upper-limb prosthetics, and other areas. An improved understanding of tactile signals gathered by the human hand is also needed in order to guide the design of tactile experiences elicited by emerging products and devices.

5.1.1 Related Research

Many tactile sensing devices are described in the literature, including several that have been applied to wearable and in vivo tactile sensing (Table 5.1). However, the task of capturing transient mechanical signals in tissues of the entire hand presents difficult challenges due to high density of tactile sensors, the distributed nature of the skin, the relatively large frequency range concerned, the large dynamic range of displacements involved (spanning 5 orders of magnitude), and the kinematic complexity of hand movements that are involved in many hand activities. Techniques that have been previously investigated for capturing transient or vibratory signals in spatially distributed regions of soft tissues of the body include large area vibrometry methods, such as ultrasonic Doppler vibrometry [31, 41], laser scanning vibrometry [33, 52], line laser sensors [19], high-speed cameras [1, 43], among others. These typically require the hand to remain stationary, making it difficult to collect tactile data during hand activities, as we aim to accomplish.

Many wearable sensing devices have also been previously investigated, including skin-wearable accelerometer arrays, such as the devices employed in our earlier work [36, 38], which are too kinematically constrained to admit most normal hand actions, partly motivating the device presented here. Others consist of sensor arrays with low spatial resolution that have not been designed to transduce propagating signals throughout the hand [10, 12, 42, 46, 49]. Another goal of prior research has been to design skin-like tactile sensors for strain [3, 6, 7, 13, 51, 53], pressure [2, 9, 21–23, 29, 30, 32, 47, 55], or similar mechanical signals. Due to the sensing principles used, these devices are unable to capture tactile signals with the large bandwidth required for capturing distributed vibrations in the hand.

Table 5.1 Sensing distributed skin vibration: Selected methods described in the literature

Reference	Sensor Technology	Wearable	Position	Data Type	Limitations
Rossi et al. [33]	Laser scanning vibrometer	No	Noncontact	Velocity scalar	Requires immobilizing hand
Seo et al. [37]	3D laser scan	No	Noncontact	Position vector	Requires immobilization Low frame rate
Sakai et al. [35]	Optical coherence tomography	No	Noncontact	Tomography	Limited spatial measurement range Requires immobilization
Tanaka et al. [43]	High-speed camera	No	Noncontact	Displacement scalar	Requires immobilization
Gerhardt et al. [8]	Microscope-video camera	No	Noncontact	Image	Small field of view Low frame rate
Shirkovskiy et al. [40]	Airborne ultrasound vibrometry	No	Noncontact	Vibrometry image	Requires spatial interpolation Requires immobilization
Sikdar et al. [41]	Ultrasonic Doppler vibrometry	No	Contact	Vibrometry image	Limited spatial measurement range Large sensor mass
Sofia et al. [42]	Accelerometer array	Yes	Contact	Acceleration vector	Low spatial resolution Limited coverage
Shao et al. [38]	Accelerometer array	Yes	Contact	Acceleration vector	Tethered Does not admit large motion
Tanaka et al. [44, 45]	PVDF film	Yes	Contact	Acceleration scalar	Very low spatial resolution Physical units unclear
Harrison et al. [10]	Piezoelectric cantilever array	Yes	Contact	Acceleration scalar	No spatial coverage Low frequency bandwidth
Our sensing method	Accelerometer array	Yes	Contact	Acceleration vector	Wired connection required

5.1.2 Contents and Contributions

In this article, we present a new 126-channel wearable sensing instrument designed to investigate information content in mechanical signals propagating in the human hand, mirroring the capabilities of the network of vibration-sensitive mechanorecep-tors that are widely distributed in hand tissues. We designed the device to accommo-date normal hand movements, a large frequency bandwidth, and spatial resolution sufficient to capture propagating mechanical waves in the regime relevant to touch perception. The lightweight and flexibility of the device allow it to be adapted to a variety of hand sizes and shapes, and to be used on either side of the hand (i.e., palmar or dorsal face). We designed a custom field programmable gate array (FPGA)-based wearable DAQ platform and communication protocol in order to interface with the accelerometer array and transfer a large amount of resulting data to a computer in real time. We demonstrate the ability of this device to capture wave propagation across the entire hand, and also demonstrate the methods for reconstructing these signals from measurements, yielding new insight into distributed information captured by the sense of touch, and a new model for tactile sensing via large area remote vibration sensing in the hand.

The wearable sensing instrument presented here has many potential applications. These include:

- Wearable systems for characterizing tactile signals felt by the hand during interac-tion with objects or haptic displays. In this application, our sensing instrument may be used to quantitatively characterize tactile experiences for applications in product design, material selection, and haptic interface characterization. The mechanical measurements may be integrated with a software tool that is able to predict the results of user testing. Such a system may replace time-intensive product testing protocols or perceptual experiments.
- Wearable devices for human–computer interaction capture interactions between the hand and touched objects. In this application, the accelerometers in our device may be used for both contact sensing and kinematic pose estimation (hand track-ing). Systems of this nature may be used in many applications for interactive computing and augmented and virtual reality. Results from the high-resolution device presented here may also be used in order to optimize sensor selection and placement for a lower complexity and lower cost device suited for many practical applications.
- Tactile sensing for robotic hands or other end effectors [4, 11, 28, 54]. Such an application may also involve the integration of this device with a skin-like covering that provides mechanical coupling analogous to that in the human hand. In this application, our device may be used to endow a robot with the ability to capture rich tactile information via a distributed array of sensors that need not be positioned on the contacting surface of the robotic limb, improving the design of the robot.
- Measurement systems for neuroscience studies. Such an application would lever-age the ability of our device to capture tactile signals from the whole hand. The values of such methods are demonstrated in our prior research [38]. An important

advantage of the new sensing methods described here is their compatibility with a large variety of hand movements, behaviors, and interactions (Sect. III).

5.2 Tactile Sensor Array Design

We designed a sensing instrument to capture the distributed response of the skin across the entire hand to multi-finger interactions of the naturally behaving hand, mirroring the vibration-sensing capabilities of the biological hand. Elasticity couples skin motion at different locations, implying that it is only necessary to sample at an array of points with a nominal density determined by the spatial Nyquist frequency. At the oscillation frequencies of interest (less 1000 Hz), the wavelength is determined by the wave speed, which depends on the propagation regime (i.e., shear, surface, or compression) and tissue properties. Prior research suggests that tactile signals travel primarily as surface or shear elastic waves in soft tissues. Their wavelengths λ are approximated by

$$\lambda = \frac{v_s}{f}, \quad v_s = \frac{E}{2\rho(1+\mu)},$$

where v_s is the shear wave speed ($\approx$ 4.4–17.5 m/s, [25]), E is the elastic modulus ($\approx$ 0.13 MPa, [20]), ρ is the density ($\approx$ 1.02 g/cm^3, [24]), and μ is Poisson's ratio ($\approx$ 0.5, [24]). In the regime relevant to tactile perception, $f < 1000$ Hz, and the wavelength λ is greater than 1 cm. This motivates sampling the surface motion of the skin at a sparse array of points, as in the presented instrument.

5.2.1 Overview of Electronics Design

To achieve this, we designed the system to integrate an array of 42 multi-axis accelerometers with wide frequency bandwidth (Fig. 5.1). We capture data at high speed via an FPGA-based multi-channel DAQ board, with the firmware on the FPGA and software on a PC, which may be a desktop or laptop system, or an embedded device. The accelerometer array is readily attached to the skin, capturing skin vibration (or other facets of vector acceleration, such as motion and gravity) over extended areas (Fig. 5.2). The DAQ board receives accelerometer readings through 23 I²C buses and then transmits them to a PC or embedded system through a universal serial bus (USB) cable. The measurement process is controlled via USB.

   In order to ensure wearability, the system was designed to be lightweight. The total weight of the sensor array, including accelerometers and flex PCBs, was 4.1 g. The flex PCB has a mass of less than 10 g. The FPGA board has a mass of 19.0 g, and dimensions (87.0 × 50.5 mm) small enough so that it can be easily mounted on the arm by using a 3D printed, conforming bracket.

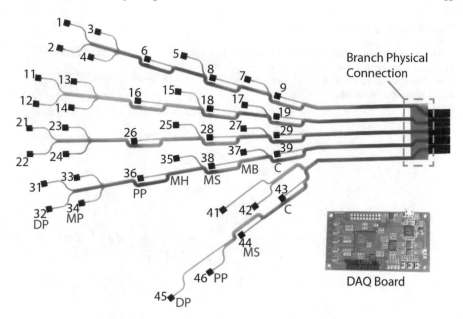

Fig. 5.1 Top view of the sensor array mounted on a flat surface, together with the FPGA-based DAQ board, is shown at the bottom. Each accelerometer is uniquely numbered. Based on the hand anatomy, the array is divided into five branches, in correspondence to digits I to V. Sensor locations were specified in relation to anatomical features, on phalanges (DP, MP, PP), metacarpals (MH, MS, MB), and carpals (C), marked in red. All five branches have a dedicated electronic connection to the DAQ board

5.2.2 Accelerometer Array

The sensor array contains 42 three-axis miniature accelerometers (Model LIS3DSH, ST Microelectronics). Each axis of the accelerometer has a selectable measurement range from ± 2 to ± 16 g, with corresponding sensitivity ranging from 0.06 to 0.73 mg. The maximum output data rate is 1.6 kHz, ensuring that the bandwidth (up 800 Hz) approximates the sensitivity range of the ensemble of low- and high-frequency-sensitive mechanoreceptors in the hand. The small package of the accelerometer ensures a contact area less than 5.0×5.0 mm. Acceleration recordings are digitized and exported to the DAQ board via I^{2}C buses.

Mechanical design

The accelerometer array and DAQ system are connected via a set of five flexible printed circuit board (PCB)s. The shape of the PCBs is ergonomically designed with bowed regions that maximize flexibility, prevent cable contact interference during hand interactions, and admit unconstrained motion in each of the segments of the

Fig. 5.2 Structure of the wearable sensor and overview of the instrument attached to the back of the hand. Each sensor is attached to the skin via a prosthetic adhesive

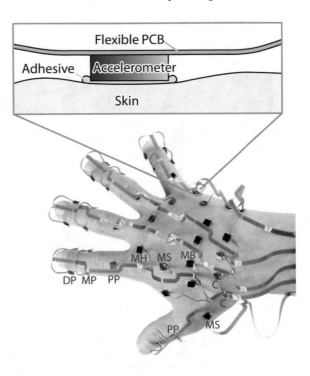

hand. The same features ensured that the sensing array can be worn on hands of various sizes (Fig. 5.3). The dimensions of the PCBs were determined based on studies of hand anthropometry [14, 48].

The sensors can be distributed over skin areas throughout the dorsal (back) surface of the hand (as indicated by the red squares in Fig. 5.3), at locations where contact rarely occurs during manual interactions, but the device can also be mounted on the palmar surface should measurement conditions dictate. Standard temporary prosthetic glue (Pros-Aide, FXWarehouse) is used to attach the accelerometers to the skin, to ensure a consistent flexible bond over a small contact patch. Thin prosthetic tape (Topstick, Vapon, Inc.) is a substitute for adhesives, enabling more rapid attachment. The accelerometer array can also be integrated into a glove to increase the flexibility of putting on and removing the sensors.

The PCB layout is designed based on hand anatomy and motion considerations (Table 5.2), accounting for movements of all five digits and the back (dorsum) of the hand, including the following anatomical regions: distal, intermediate, and proximal phalanges, the metacarpal and carpal areas (Fig. 5.1). To avoid artifacts, we avoided positioning the sensors on joints, where the skin surface deforms the most during digit movement. Instead, sensors are positioned on or between bones. This also reduced interference introduced by movement and ensured consistent attachment of the sensors. The device is divided into five independent PCBs, or main branches, each corresponding to one digit. The cable morphology was revised through sev-

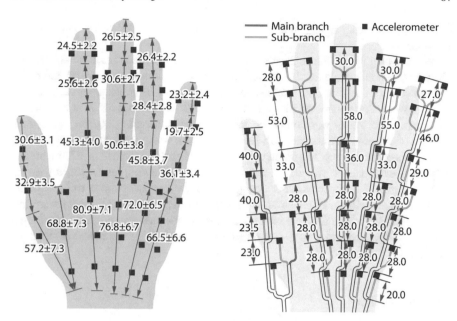

Fig. 5.3 Design of the sensor array was based on hand anthropometry (mm). Left: Statistics (mean ± standard deviation) of dorsal hand anthropometric dimensions of 69 females and 70 males, with age ranging from 18 to 68 [48]. Red squares indicated the example locations of the sensors mounted on the hand. Right: Designed dimension of the flexible PCBs. All sub-branch cables (blue) have a length of 20 mm, in order to adapt the variance of hand sizes and admit unconstrained motion of the hand

Table 5.2 Hand anatomical locations of the sensor

Anatomical location	Accelerometer number
Distal phalanges (DP)	1, 2, 11, 12, 21, 22, 31, 32, 45
Middle phalanges (MP)	3, 4, 13, 14, 23, 24, 33, 34
Proximal phalanges (PP)	6, 16, 26, 36, 46
Metacarpals head (MH)	5, 15, 25, 35
Metacarpals shaft (MS)	8, 18, 28, 38, 44
Metacarpals base (MB)	7, 17, 27, 37
Carpals (C)	9, 19, 29, 39, 43
Between MS of digits I, II	41
Between MS and MB of digit II	42

eral iterations using paper mockups so as to impose minimal constraint to finger movements.

In early prototypes, the sensors were placed directly on one main branch cables. However, residual cable tension limited the range of motion of the hand and could induce contact stress at the accelerometer attachment. In subsequent designs, the sensors were then connected to the FPGA through flexible cables with a branched structure. Each sensor was connected to a main branch cable (3.0 mm width, connection of all sensors) through a sub-branch (1.3 mm width, connected to one sensor only). When the sensors are affixed to the hand, the sub-branch cables lift the branch cable above the hand to prevent contact with the skin, allowing for flexion and extension of digits. Those cables were designed to have a curved shape to have better flexibility and facilitate PCB layout. Unlike the distal phalanges (DP) and the middle phalanges (MP) where the sensors were placed on the side, the proximal phalanges (PP) had a sensor on top due to space limitations on their side in most normal hand interactions.

Electrical design

The sensors are divided into five branches, each connected to a dedicated port on the DAQ board. All of them operate independently. The system remains functional if any of the five branches is removed, as can be accomplished by trimming the connective area of the PCB. This feature was designed to accommodate demands for task-specific or localized measurement, for example, during interaction of a single digit. Each branch has independent circuit connection with the FPGA board, with five independent I^{2}C buses, with power and ground connections. All five branches are physically connected at the end of the PCB, near the connectors, and can be separated easily by cutting through a nonfunctional area between branch circuits (Fig. 5.1). This makes it possible to deploy the device in configurations that need not include all fingers, or involve other usages not presented here, such as simultaneous sensing on both faces of selected digits, on other regions of the limb or body.

The PCB design accommodates sensor configurations that vary depending on the size and shape of the hand on which they are mounted, with locations on the hand surface and relative orientations that are registered to the anatomical features of the hand. For any sensor placement and fixed posture of the hand, the configuration of each of the N_s sensors is described by a set of spatial positions and orientations, in three dimensions, which can be expressed as pose matrices T_i given by

$$T_i = \begin{pmatrix} R_i & \mathbf{p}_i \\ 0_{1\times3} & 1 \end{pmatrix} \tag{5.1}$$

with

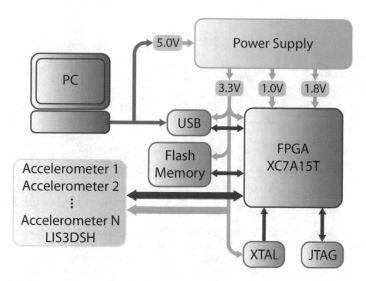

Fig. 5.4 Overview of the wearable sensing system. N is the number of accelerometers that are connected to the FPGA through I²C buses on flexible PCBs. N∈{6, 9, 15, 18, 24, 27, 33, 36, 42}. The data is streamed in real time to a PC or embedded device via a USB connection, which also supplies the power for the instrument. The green and red arrows indicate power connection and communication signal lines, respectively. Power and communication are both handled over USB, as depicted as blue arrows

$$\mathbf{p}_i = (p_{i,x}, \ p_{i,y}, \ p_{i,z})^\top, \tag{5.2}$$

$$R_i \in SO(3), \tag{5.3}$$

$$i = 1, 2, \ldots, N_s, \tag{5.4}$$

where $\mathbf{p}_i$ are the coordinates of the ith sensor relative to a fixed coordinate system of the spatial environment, and R_i is a 3×3 rotation matrix, which specifies an orthonormal frame, describing the orientation of the ith sensor relative to this environment. Both $\mathbf{p}_i$ and R_i depend on the kinematic pose of the hand, and the relative position and orientation of different sensors can change depending on the hand position orientation, and the pose of the fingers.

In Sect. 5.3.3, we describe how to map the sensor configuration onto a standard reference hand, for the purpose of integrating and analyzing tactile signals across the geometric surface of the hand.

5.2.3 FPGA-Based Multi-channel Data Acquisition Board

We designed an FPGA board (model XC7A75T, Artix-7 series, Xilinx inc.) to acquire data from the 42 accelerometers. An overview of the system is shown in Fig. 5.4.

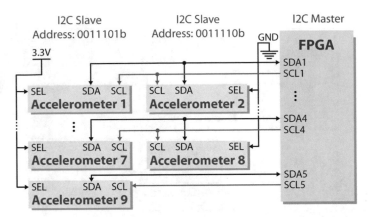

Fig. 5.5 I^2C communication diagram between the FPGA (master) and accelerometers (slaves). Each I^2C bus is shared by one pair of accelerometers that communicate with the FPGA alternatively. SCL: Serial clock line. FPGA outputs a clock of 1.6 MHz (maximum) to each pair of sensors. SDA: Serial data line. SEL: I^2C address selection. For each pair of accelerometers, the one with this pin pulled up to 3.3V has I^2C address of 0011101b, while the pin of the other one is grounded to have I^2C address of 0011110b. Power supply lines for accelerometers are not displayed in this diagram. Only one branch of the flexible PCB, consisting of accelerometer 1 to 9, is shown in this diagram

In order to capture data from the ensemble of sensors in real time, the FPGA was programmed to interface with the accelerometers through 23 I^2C bus connections. Those buses are used to transmit sensor recordings to the FPGA in a parallel manner. They are divided into five branches, corresponding to sensor mounting locations on the five digits. The branches on digit II to V each include 5 I^2C buses, while the branch on digit I contains 3 buses. The bus branch on digit V contains accelerometers 1 to 9, with each I^2C bus communicating with two accelerometers except the last one, which is connected to accelerometer 9 only (Fig. 5.5). Similarly, branches on digits IV, III, and II contain accelerometer 11–19, 21–29, and 31–39, separately. The numbers 10, 20, 30, and 40 were skipped as they have no I^2C connection. The branch on digit I contains three I^2C buses, connecting accelerometer 41 to 46. All branches have isolated circuit routes and interface with the FPGA via separate AXT6 pitch connectors on the board, which enables the separation of any branch from the others. The board includes a JTAG module (IEEE 1149.7) that makes it possible to interface the board with a PC via JTAG-SMT2, for programming and debugging purposes. An N25Q256A flash memory module enables the DAQ program to boot and operate without JTAG connection with the PC.

Our firmware design enables data to be streamed to the PC in real time through a USB cable, which also supplies power to the FPGA board. We have interfaced the device with a desktop PC and with embedded computers running Windows and Linux operating systems.

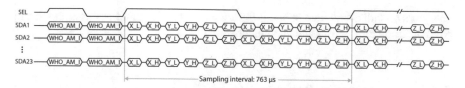

Fig. 5.6 Data sampling protocol of the wearable sensor array through 23 I²C buses. WHO_AM_I register of each accelerometer is checked before each measurement. Then, data is sampled from accelerometer SEL pull-up and pull-down groups alternatively. For a sample of an accelerometer, X, Y, and Z-axis data is acquired sequentially, each with the low byte (L) followed by the high byte (H)

5.2.4 Communication Protocol and Data Storage

Communications between the accelerometers and FPGA follow a data sampling protocol shown in Fig. 5.6. In order to maximize the number of sensors that can be interfaced, we index sensors on each I²C bus via the SEL bit. For any pair of accelerometers connected to the same I²C bus, the data are in an alternating fashion. Each is associated with a different slave address indexed via the SEL bit. Accelerometers with SEL high (pin voltage 3.3 V) are associated with an I²C address with SEL = 1, while those with SEL pin grounded have an address with SEL = 0. Data are acquired in interleaved fashion, first from the 23 odd-numbered accelerometers (SEL = 1), then from the remaining 19 even-numbered accelerometers (SEL = 0).

One sample of data from an accelerometer returns data (a_x, a_y, a_z) corresponding to the Cartesian components of acceleration in the local frame given by the orientation of the accelerometer (as we demonstrate below, this data can be interpreted for analysis even if the orientation frame is unknown). Before reading a sample, data availability of all accelerometers is verified via the status registers, which are designed to synchronize data transmission among the I²C communication buses. Although the latter have independent clock lines, these lines are all synchronized and clocked to 1.6 MHz. After accounting for communication overhead, the sampling frequency is 1310 Hz. Data from all sensors in the array (including both odd and even indexed sensors) are captured within one period at this rate. Accelerometer data received by the FPGA are transferred to the PC via USB and stored in a binary file in real time.

5.3 Validation Experiments and Results

We investigated the performance of the sensing system by examining data captured by the sensors, and by investigating the applicability of this device to capturing distributed vibrations in the skin, including vibration signatures produced through touch contact. We assessed the device in both bench-top experiments that directly probed the performance of the sensors in the array and through testing as the device was worn on individual human hands during a variety of manual gestures and during passive stimulation with controlled mechanical inputs.

5.3.1 Sensor Array Performance

The accelerometers produce vector signals $\mathbf{a}_i(t) = (a_{i,x}, a_{i,y}, a_{i,z})$, where i indexes the sensor and $t = 0, \tau, 2\tau, \ldots, N\tau$ indexes time. The sample period $\tau = 1/f_s = 0.76$ milliseconds. In order to integrate information from the ensemble of these vector measurements for the analysis, the positions $\mathbf{p}_i$ and orientations R_i of each sensor should be known. While the former is known relative to anatomical landmarks, the latter is more challenging to acquire, as the orientation depends on the pose of the hand and orientation of the skin. To address this, we developed a principal components analysis-based method for computing an orientation-invariant scalar value that captures the most salient available at each sensor. The method is to project the vector sensor signal onto an estimate of its instantaneous principal component. This operation is linear, avoiding nonlinear artifacts that would be introduced by computing a magnitude, and preserves crucial phase information. Using data X_i captured from each sensor over a sliding window of hundreds of samples, we first compute the eigenvector $\mathbf{w}$ of the covariance matrix $\mathbf{X}_i = \{\mathbf{a}_i(0)\ \mathbf{a}_i(\tau)\ \cdots\ \mathbf{a}_i(N\tau)\}$ with the largest eigenvalue λ, where N is the number of measurements (typically, $100 < N < 1000$, corresponding to a small fraction of a second). By abuse of notation, we refer to this scalar value as $a_i(t)$, where

$$a_i(t) = \mathbf{a}_i^T \mathbf{w}, \quad \mathbf{w} = \arg\max \left\{ \frac{\mathbf{w}^T \mathbf{X}^T \mathbf{X} \mathbf{w}}{\mathbf{w}^T \mathbf{w}} \right\}. \tag{5.5}$$

No temporal smoothing or filtering is performed. With this approach, the projected scalar acceleration value computed from each sensor signal depends on the known position, but not the unknown orientation, of the sensor. It also depends on the applied stimulus, which can differentially excite vector skin motion at different locations.

We tested the device performance when worn on a human hand (Fig. 5.7) as the hand was stimulated by 100 Hz sinusoidal stimulus applied to the tip of digit II of the subject using a controlled electromechanical stimulator (Model 4810, Brüel and Kjær). The hand was supported with digit II extending to contact the steel probe of the stimulator. The output of the ensemble of sensors was captured, and scalar acceleration signals computed for each accelerometer. A 200 ms segment of the values captured from one branch of the accelerometer array, representing one-fifth of the accelerometers, is shown in Fig. 5.7. Signals of the branch that correspond to digit II are shown in the figure. The signals measured at distributed points in the skin retain a highly sinusoidal waveform, reflecting the fact that signal propagation in hand tissues is highly linear at these frequencies. The amplitude decreased with increasing distance, reflecting energetic losses due to damping in hand tissues (Figs. 5.8 and 5.9).

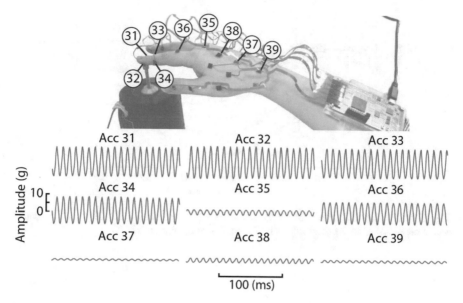

Fig. 5.7 A 100 Hz sinusoidal stimulus was applied to the tip of digit II of the subject. Acceleration in the direction with the highest vibration amplitude $s_i(t)$ is shown for accelerometer 31–39 (digit II branch)

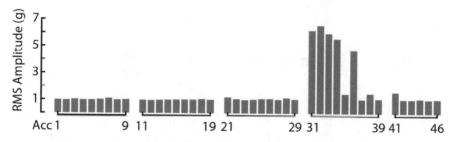

Fig. 5.8 RMS acceleration amplitude measured at all locations when 100 Hz sinusoidal stimulus was applied to the tip of digit II

We also validated the measurements captured with individual sensors in our device using an accurate, noncontact laser Doppler vibrometer (Model PDV-100, Polytec). The LDV had a sampling frequency of 22 kHz. The measurement range was set to be 100 mm/s. The maximum acceleration was 13800 m/s^2, and the resolution was 0.02 $\frac{\mu m}{s}/\sqrt{Hz}$. A single accelerometer of the device was attached to the probe of the exciter vibrating (100 Hz sinusoidal signal) along the Z-axis of the accelerometer. A thin reflective tape was attached to the top of the accelerometer board, where the LDV was directed. Readings of all accelerometers were recorded through the device. We compared the output of the sensor that was attached to the probe with the signal from the LDV. The recording of the LDV was low-pass filtered 1000 Hz to better match the bandwidth of the sensor array, and differentiated to obtain acceleration, which

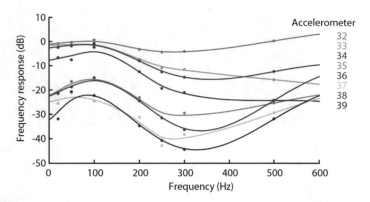

Fig. 5.9 Location-dependent frequency response of accelerometer 32–39 with respect to accelerometer 31. (100 Hz sinusoidal stimulus applied at tip of digit II)

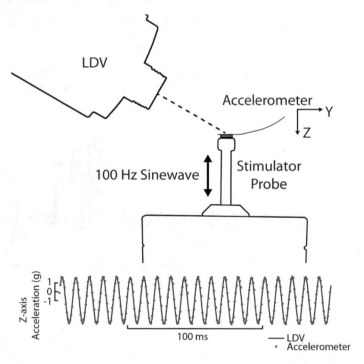

Fig. 5.10 A signal waveform of one accelerometer mounted on the tip of an actuator, with the Z-axis pointing along the actuator's vibration direction. The actuator output is 100 Hz sinusoid. Comparison with LDV acceleration measurement along the Z-axis direction was made

was compared with the $a_z(t)$, the accelerometer measurement component along the Z-axis (Fig. 5.10). The results indicate that measurements of the accelerometer and LDV were in close agreement.

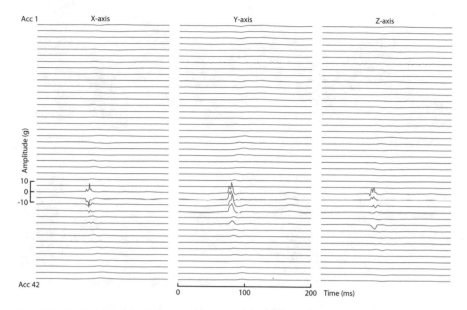

Fig. 5.11 X, Y, and Z-axis signal waveforms for all 42 accelerometers (one per row, numbered from top to bottom) during tapping digit II

The current consumption is approximately 9.5 mA for the entire array (225 μA per sensor in active mode, or 2 μA in standby mode), at a typical operating voltage of 2.5 V. Thus, the power consumption is low, enabling operation via a compact battery or a standard USB port. For the same reason, no significant temperature increase is observed even during long-term operation (≳ 1 h). The sensors produce stable output up to operating temperatures of 85 °C, and no temperature-dependent effects are observed in the applications described here.

5.3.2 Wearable Sensing Experiments

We further evaluated the ability of the system to accommodate natural movements and touch contacts of the hand using a representative variety of hand actions, for the purpose of capturing distributed vibrations (Fig. 5.11) excited by skin–object contact during everyday interactions (Fig. 5.12). Here, we emphasize the ergonomic aspects of the device. This shows that with the sensor array configured on the hand as shown, a wide variety of manual interactions is possible, without constraints on the motion of the hand or digits.

In addition, we evaluated the similarity between tactile signals elicited by different gestures (Fig. 5.13), based on measurements of $N_s = 42$ accelerometers. To compare the signals elicited by one gesture $A(t) = \{a_1(t)\ a_2(t)\ \cdots\ a_{N_s}(t)\}$ to the signals

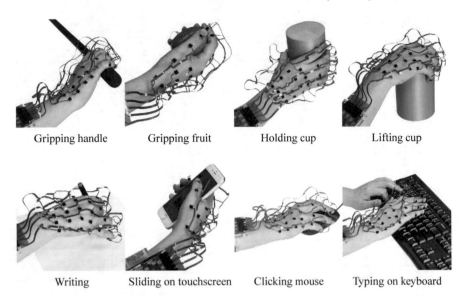

Gripping handle　　　Gripping fruit　　　Holding cup　　　Lifting cup

Writing　　　Sliding on touchscreen　　　Clicking mouse　　　Typing on keyboard

Fig. 5.12 Capturing tactile signals when performing natural hand gestures. There were no contact between cables and the skin

Fig. 5.13 Similarity computed from sum of maximum correlation between tactile signals elicited by different gestures

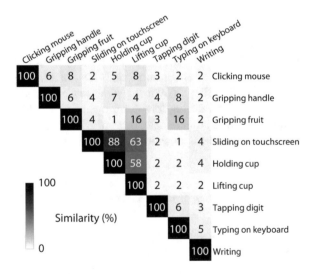

	Clicking mouse	Gripping handle	Gripping fruit	Sliding on touchscreen	Holding cup	Lifting cup	Tapping digit	Typing on keyboard	Writing	
	100	6	8	2	5	8	3	2	2	Clicking mouse
		100	6	4	7	4	4	8	2	Gripping handle
			100	4	1	16	3	16	2	Gripping fruit
				100	88	63	2	1	4	Sliding on touchscreen
					100	58	2	2	4	Holding cup
						100	2	2	2	Lifting cup
							100	6	3	Tapping digit
								100	5	Typing on keyboard
									100	Writing

100

Similarity (%)

0

elicited by another gesture $\bar{A}(t) = \{\bar{a}_1(t)\ \bar{a}_2(t)\ \cdots\ \bar{a}_{N_s}(t)\}$, we computed a similarity score given by

$$S(A, \bar{A}) = \frac{\sum_{i=1}^{N_s} \max_\tau(|\int_t a_i(t)\bar{a}_i(t+\tau)dt|)/\sigma_{a_i}\sigma_{\bar{a}_i}}{\sum_{i=1}^{N_s} \int_t \bar{a}_i(t)^2 dt/\sigma_{\bar{a}_i}^2}, \quad (5.6)$$

where σ_x is the standard deviation of a signal x. The results indicate how tactile signals elicited by different actions are highly distinct. The tactile signals that were most similar (e.g., grasping a cup or sliding fingers on a hand-held touch screen) were those that engaged similar combinations of digits in similar configurations.

5.3.3 Reconstructing Tactile Wave of Natural Stimuli

As explained in Section II, elasticity couples skin motion at different locations. The spatial wavelengths of signals excited in the range of frequencies relevant to tactile perception are relatively long ($\lambda > 1$ cm), allowing a sparse Nyquist sampling approach. Here, we demonstrate how to estimate the spatiotemporal skin motion across the hand from accelerometer measurements, by mapping them onto a standard 3D hand reconstruction. Each accelerometer signal $\mathbf{a}_i(t)$ is associated with a position $\mathbf{p}_i$. We associate these locations to the coordinates of the 3D hand model via anatomical registration of the anatomically known sensor locations. The hand model consists of a point cloud obtained from a high-resolution 3D scanner (Model Eva, Artec3D). Using the model, we compute the geodesic distance $d(\mathbf{p}_i, \mathbf{p})$ between sensor positions $\mathbf{p}_i$ and arbitrary surface points $\mathbf{p}$, using a shortest path algorithm. We estimated the acceleration $\mathbf{a}(\mathbf{p}, t)$ at arbitrary hand surface points using a physiologically informed distance weighting (Fig. 3.5) [25, 38]:

$$\mathbf{a}(\mathbf{p}, t) = \frac{\sum_{i=1}^{42} f(\phi(\mathbf{p}, \mathbf{p}_i))\mathbf{a}_i(t)}{\sum_{i=1}^{42} f(\phi(\mathbf{p}, \mathbf{p}_i))}, \tag{5.7}$$

$$\phi(\mathbf{p}, \mathbf{p}_i) = \frac{17}{d(\mathbf{p}, \mathbf{p}_i) + \alpha} - C, \tag{5.8}$$

$$f(\phi) = \begin{cases} \phi, & \phi \geq 0 \\ 0, & \phi < 0. \end{cases} \tag{5.9}$$

Here, $f(\phi)$ is a rectifier (replacing all negative values with zeros). We evaluated these equations using values $\alpha = 25.5$ mm and $C = 8.7 \times 10^{-2}$ that we selected based on previously published measurements [25]. $\phi(\mathbf{p}, \mathbf{p}_i)$ is used to compute the vibration amplitude at points $\mathbf{p}$ based on measurement of sensor $\mathbf{a}_i(t)$ at $\mathbf{p}_i$ accounting for damping with distance $d(\mathbf{p}_i, \mathbf{p})$. While we could readily accommodate hand size, via a scale parameter γ, differences in hand shape and mechanics would require further steps.

We applied this method in order to reconstruct spatiotemporal patterns of skin acceleration in the whole hand during manual interactions. Figure 5.8 illustrates that vector acceleration signals $\mathbf{a}_i(t)$ recorded for all sensors and channels for one of these gestures, involving tapping the index finger, digit II. We performed reconstructions of skin acceleration in the whole hand during manual interactions for an ensemble of different gestures. Figure 5.14 depicts the instantaneous skin acceleration associated

Fig. 5.14 Left: Photo showing the gesture that elicited the wave patterns shown. Right: Time-averaged (RMS over a 250 ms window) vibration amplitude of tapping digits I to V on a flat surface

Fig. 5.15 Left: Photo showing the gesture that elicited the wave patterns shown. Right: Grabbing and lifting a handle. Wave propagation was reconstructed on a 3D hand model. Five selected instants showing whole-hand tactile signal evolution during grabbing and lifting a handle

with each of five gestures that involved tapping each respective finger, revealing that these single-digit interactions elicited energy that readily propagated along the full extent of the respective digit, and into the rest of the hand.

The wearable sensor captures tactile signals that are elicited in the whole hand during object grasping and manipulation, as shown in Fig. 5.15. Here, grasping and lifting a handle excites skin vibration in all digits at different times, due to variations in contact timing. These vibrations readily propagate throughout the rest of the hand.

These results consist of individual measurement trials and reveal how localized contact with the skin excites substantial vibrations that are transmitted to sensors distributed over large areas of the hand, with time delays that reflect the mechanics of wave transmission. The patterns of these vibrations vary with the gestures, interactions, and objects that are involved. Recent neuroscience research shows how these propagating vibrations support human perception and action [5, 25, 38]. The results presented here provide an empirical demonstration of how such information can be captured via a sparse sensor array—one that coarsely approximates the vibration-sensitive tactile channel in the human hand.

5.3.4   Measuring Both Sides of the Hand

During measurements involving human grasping, it can be useful to avoid occluding the palmar surface of the hand, which could otherwise interfere with human grasping and perception, and introduce contact artifacts. In other applications, such a sensor array can be worn on the palmar or nonpalmar (dorsal surface). Due to the mechanical

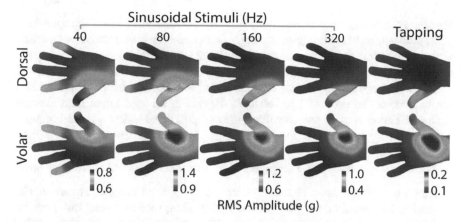

Fig. 5.16 Due to mechanical coupling in the hand, and the large wavelengths ($\lambda > 1$ cm[38]) of the propagating signals, relatively similar vibration patterns occur in both sides of the hand (dorsal and palmar surfaces). RMS acceleration magnitude of vibrations elicited by sinusoidal stimulation on the palmar surface at the base of the thumb (thenar eminence). Stimuli were applied at different vibration frequencies, or via discrete tapping

coupling between hand tissues, and the large wavelengths ($\lambda > 1$ cm [38]) of the propagating signals, relatively similar vibration patterns occur in both sides of the hand (Fig. 5.16).

5.4 Conclusion

In this chapter, inspired by human sensing, anatomy, and tissue biomechanics, we present a wearable tactile sensing array for large area remote vibration sensing in the whole hand. The system comprises a 126-channel, wide bandwidth sensing array based on an ergonomically designed, flexible apparatus integrating 42 discrete three-axis accelerometers, and capable of capturing the propagation of vibrations, in the form of viscoelastic waves. These waves propagate efficiently in hand tissues, distributing mechanical signatures of touch to widespread hand areas and sensory resources.

The electronic design is based on a custom FPGA-based data acquisition system and sampling across 23 discrete I²C networks, which ensures that the data across the array can be captured with high temporal fidelity and resolution, with an effective sampling rate 1310 Hz. By carefully designing the physical configuration and ergonomics of this wearable sensor array, we ensure that it can be used to accurately capture contact-elicited tactile signals during a variety of natural interactions, including touch contact, grasping, and manipulation of objects. Experiments show that the individual sensors transduce skin acceleration with high temporal resolution. We also showed that the device can accommodate a wide range of hand motions. We also introduce physiologically informed methods for 3D reconstruction of the spatiotemporal propagation of tactile signals throughout the hand. Reconstruction was possible with a sparse array due to the comparatively large wavelength of the propagating tactile signals ($\lambda > 1$ cm). The results show how the captured spatiotemporal signals vary with the performed actions, objects, and skin locations. In some applications, the number of sensors could be reduced, depending on application requirements, such as the number of gestures involved, the computational task (e.g., discriminating a few vs. many gestures), or the subset of hand regions involved (e.g., one finger vs. entire hand).

Such sensing methods hold the potential to contribute to several application areas:

- They can enable sensing instruments for product design that can capture whole-hand tactile signals during manual activities using measurements that preserve detailed temporal and spatial signal attributes reflecting the properties of touched objects and the manner in which they are touched. Capturing such data is less time-intensive than widely used human subject methods for product testing or perception evaluation. Thus, these methods hold potential for the design and assessment of haptic interaction with commercial products.
- Our results furnish sensing methods that may improve robotic hands, by endowing them with the ability to capture tactile information, such as transient contact signatures, via a distributed sensor network that shares properties with the vibration-sensitive resources in the human hand (see Introduction). In humans, these resources are crucial for enabling sensorimotor grasping and dextrous manipulation [15].
- Our findings may advance interactive, human-wearable technologies. The methods we demonstrate can supply design guidelines based on measurements captured in the hand during a large variety of interactions. With such measurements, a designer can optimize sensor configurations for a wearable product with fewer sensors, or can design gesture-based interactions (for interactive computing, virtual and augmented reality) that are more readily discriminable. The signals captured by our system preserve and reflect the differences between the gestures involved and the marked differences between signals captured at different skin locations.
- Such sensing systems also hold promise for tactile neuroscience research. A major gap in existing research methods is that it is rarely possible to capture the mechanical stimuli felt by the hand, which are analogous to images captured by the eye. Our results show how detailed information about tactile stimuli felt in the naturally behaving hand may be captured in vivo. Several new studies in our group and labs of our collaborators are applying the methods presented here to address important questions in the neuroscience of touch perception and interaction.

Recent research has elucidated the remarkable abilities of the human hand to perform touch sensing at a distance, using remotely propagating viscoelastic waves that are excited via contact interactions with objects and surfaces. The exquisite biological sensory apparatus of the human hand leverages these propagating tactile signals in order to enable a large variety of perceptual and manipulation tasks. This research enables new methods for scientific inquiry, including studies of how tactile sensing in the whole hand relies on both mechanics and neural transduction to shaping tactile inputs that are the basis of touch perception and prehensile interaction. It also provides a unique model for artificial touch sensing, in which contact interactions are encoded via a multitude of vibration sensors positioned remotely from the contact interface, and are synergetically combined through mechanical coupling. The instrument presented here also enables new methods for tactile sensing in wearable applications of interactive computing, and provides new quantitative methods for product design and usability that leverage these new tools for characterizing tactile signals that are felt during interaction with newly designed products or interfaces. We anticipate that future devices inspired by the system presented here will facilitate applications of this approach in robotics, prosthetics, consumer electronics, wearable computing, augmented and virtual reality, health care, and many other applications.

References

1. Allen, R.D.: High speed video distribution and manufacturing system (1999). US Patent 5,909,638
2. Chortos, A., Liu, J., Bao, Z.: Pursuing prosthetic electronic skin. Nat. Mater. **15**(9), 937 (2016)
3. Cooper, C.B., Arutselvan, K., Liu, Y., Armstrong, D., Lin, Y., Khan, M.R., Genzer, J., Dickey, M.D.: Stretchable capacitive sensors of torsion, strain, and touch using double helix liquid metal fibers. Adv. Func. Mater. **27**(20), 1605630 (2017)
4. Dahiya, R.S., Metta, G., Valle, M., Sandini, G.: Tactile sensing-from humans to humanoids. IEEE Trans. Rob. **26**(1), 1–20 (2010)
5. Delhaye, B., Hayward, V., Lefèvre, P., Thonnard, J.L.: Texture-induced vibrations in the forearm during tactile exploration. Front. Behav. Neurosci. **6**, 37 (2012)
6. Dickey, M.D.: Stretchable and soft electronics using liquid metals. Adv. Mater. **29**(27), 1606425 (2017)
7. Do, T.N., Visell, Y.: Stretchable, twisted conductive microtubules for wearable computing, robotics, electronics, and healthcare. Sci. Rep. **7**(1), 1753 (2017)
8. Gerhardt, L.C., Schmidt, J., Sanz-Herrera, J., Baaijens, F., Ansari, T., Peters, G., Oomens, C.: A novel method for visualising and quantifying through-plane skin layer deformations. J. Mech. Behav. Biomed. Mater. **14**, 199–207 (2012)
9. Gong, S., Schwalb, W., Wang, Y., Chen, Y., Tang, Y., Si, J., Shirinzadeh, B., Cheng, W.: A wearable and highly sensitive pressure sensor with ultrathin gold nanowires. Nat. Commun. **5**, 3132 (2014)
10. Harrison, C., Tan, D., Morris, D.: Skinput: appropriating the body as an input surface. In: Proceedings of the SIGCHI Conference on Human Factors in Computing Systems, pp. 453–462. ACM (2010)
11. Howe, R.D.: Tactile sensing and control of robotic manipulation. Adv. Robot. **8**(3), 245–261 (1993)
12. Howe, R.D., Cutkosky, M.R.: Sensing skin acceleration for slip and texture perception. In: Proceedings., 1989 IEEE International Conference on Robotics and Automation, 1989, pp. 145–150. IEEE (1989)
13. Hua, Q., Sun, J., Liu, H., Bao, R., Yu, R., Zhai, J., Pan, C., Wang, Z.L.: Skin-inspired highly stretchable and conformable matrix networks for multifunctional sensing. Nat. Commun. **9**(1), 244 (2018)
14. Imrhan, S.N., Sarder, M., Mandahawi, N.: Hand anthropometry in bangladeshis living in america and comparisons with other populations. Ergonomics **52**(8), 987–998 (2009)

15. Johansson, R.S., Flanagan, J.R.: Coding and use of tactile signals from the fingertips in object manipulation tasks. Nat. Rev. Neurosci. **10**(5), 345–359 (2009)
16. Jones, L.A., Lederman, S.J.: Human Hand Function. Oxford University Press (2006)
17. Jörntell, H., Bengtsson, F., Geborek, P., Spanne, A., Terekhov, A.V., Hayward, V.: Segregation of tactile input features in neurons of the cuneate nucleus. Neuron **83**(6), 1444–1452 (2014)
18. Kandel, E.R., Schwartz, J.H., Jessell, T.M., of Biochemistry, D., Jessell, M.B.T., Siegelbaum, S., Hudspeth, A.: Principles of neural science, vol. 4. McGraw-hill New York (2000)
19. Kawahara, T., Tanaka, S., Kaneko, M.: Non-contact stiffness imager. Int. J. Robot. Res. **25**(5–6), 537–549 (2006)
20. Khatyr, F., Imberdis, C., Vescovo, P., Varchon, D., Lagarde, J.M.: Model of the viscoelastic behaviour of skin in vivo and study of anisotropy. Ski. Res. Technol. **10**(2), 96–103 (2004)
21. Li, B., Fontecchio, A.K., Visell, Y.: Mutual capacitance of liquid conductors in deformable tactile sensing arrays. Appl. Phys. Lett. **108**(1), 013502 (2016)
22. Li, B., Gao, Y., Fontecchio, A., Visell, Y.: Soft capacitive tactile sensing arrays fabricated via direct filament casting. Smart Mater. Struct. **25**(7), 075009 (2016)
23. Li, B., Shi, Y., Hu, H., Fontecchio, A., Visell, Y.: Assemblies of microfluidic channels and micropillars facilitate sensitive and compliant tactile sensing. IEEE Sens. J. **16**(24), 8908–8915 (2016)
24. Liang, X., Boppart, S.A.: Biomechanical properties of in vivo human skin from dynamic optical coherence elastography. IEEE Trans. Biomed. Eng. **57**(4), 953–959 (2010)
25. Manfredi, L.R., Baker, A.T., Elias, D.O., Dammann III, J.F., Zielinski, M.C., Polashock, V.S., Bensmaia, S.J.: The effect of surface wave propagation on neural responses to vibration in primate glabrous skin. PloS One **7**(2) (2012)
26. Manfredi, L.R., Saal, H.P., Brown, K.J., Zielinski, M.C., Dammann, J.F., III., Polashock, V.S., Bensmaia, S.J.: Natural scenes in tactile texture. J. Neurophysiol. **111**(9), 1792–1802 (2014)
27. McGlone, F., Wessberg, J., Olausson, H.: Discriminative and affective touch: sensing and feeling. Neuron **82**(4), 737–755 (2014)
28. Nicholls, H.R., Lee, M.H.: A survey of robot tactile sensing technology. Int. J. Robot. Res. **8**(3), 3–30 (1989)
29. Nie, P., Wang, R., Xu, X., Cheng, Y., Wang, X., Shi, L., Sun, J.: High-performance piezoresistive electronic skin with bionic hierarchical microstructure and microcracks. ACS Appl. Mater. Interfaces **9**(17), 14911–14919 (2017)
30. Park, J., Lee, Y., Hong, J., Ha, M., Jung, Y.D., Lim, H., Kim, S.Y., Ko, H.: Giant tunneling piezoresistance of composite elastomers with interlocked microdome arrays for ultrasensitive and multimodal electronic skins. ACS Nano **8**(5), 4689–4697 (2014)
31. Plett, M.I., Beach, K.W., Dunmire, B., Brown, K.G., Primozich, J.F., Strandness, E.: In vivo ultrasonic measurement of tissue vibration at a stenosis: a case study. Ultrasound Med. Biol. **27**(8), 1049–1058 (2001)
32. Pu, X., Liu, M., Chen, X., Sun, J., Du, C., Zhang, Y., Zhai, J., Hu, W., Wang, Z.L.: Ultra-stretchable, transparent triboelectric nanogenerator as electronic skin for biomechanical energy harvesting and tactile sensing. Sci. Adv. **3**(5), e1700015 (2017)
33. Rossi, G., Tomasini, E.: Hand-arm vibration measurement by a laser scanning vibrometer. Measurement **16**(2), 113–124 (1995)
34. Saal, H.P., Bensmaia, S.J.: Touch is a team effort: interplay of submodalities in cutaneous sensibility. Trends Neurosci. **37**(12), 689–697 (2014)
35. Sakai, S., Yamanari, M., Lim, Y., Nakagawa, N., Yasuno, Y.: In vivo evaluation of human skin anisotropy by polarization-sensitive optical coherence tomography. Biomed. Opt. Express **2**(9), 2623–2631 (2011)
36. Schäfer, H., Wells, Z., Shao, Y., Visell, Y.: Transfer properties of touch elicited waves: effect of posture and contact conditions. In: World Haptics Conference (WHC), 2017 IEEE, pp. 546–551. IEEE (2017)
37. Seo, H., Kim, S.j., Cordier, F., Choi, J., Hong, K.: Estimating dynamic skin tension lines in vivo using 3D scans. Comput.-Aided Des. **45**(2), 551–555 (2013)

38. Shao, Y., Hayward, V., Visell, Y.: Spatial patterns of cutaneous vibration during whole-hand haptic interactions. Proc. Natl. Acad. Sci. **113**(15), 4188–4193 (2016)
39. Shao, Y., Hu, H., Visell, Y.: A wearable tactile sensor array for large area remote vibration sensing in the hand. IEEE Sens. J. **20**(12), 6612–6623 (2020)
40. Shirkovskiy, P., Laurin, A., Jeger-Madiot, N., Chapelle, D., Fink, M., Ing, R.: Airborne ultrasound surface motion camera: application to seismocardiography. Appl. Phys. Lett. **112**(21), 213702 (2018)
41. Sikdar, S., Lee, J.C., Remington, J., Zhao, X.Q., Goldberg, S.L., Beach, K.W., Kim, Y.: Ultrasonic doppler vibrometry: novel method for detection of left ventricular wall vibrations caused by poststenotic coronary flow. J. Am. Soc. Echocardiogr. **20**(12), 1386–1392 (2007)
42. Sofia, K.O., Jones, L.: Mechanical and psychophysical studies of surface wave propagation during vibrotactile stimulation. IEEE Trans. Haptics **6**(3), 320–329 (2013)
43. Tanaka, N., Higashimori, M., Kaneko, M., Kao, I.: Noncontact active sensing for viscoelastic parameters of tissue with coupling effect. IEEE Trans. Biomed. Eng. **58**(3), 509–520 (2011)
44. Tanaka, Y., Nguyen, D.P., Fukuda, T., Sano, A.: Wearable skin vibration sensor using a pvdf film. In: World Haptics Conference (WHC), 2015 IEEE, pp. 146–151. IEEE (2015)
45. Tanaka, Y., Yoshida, T., Sano, A.: Practical utility of a wearable skin vibration sensor using a PVDF film. In: World Haptics Conference (WHC), 2017 IEEE, pp. 623–628. IEEE (2017)
46. Tarabini, M., Saggin, B., Scaccabarozzi, D., Moschioni, G.: The potential of micro-electromechanical accelerometers in human vibration measurements. J. Sound Vib. **331**(2), 487–499 (2012)
47. Trung, T.Q., Lee, N.E.: Flexible and stretchable physical sensor integrated platforms for wearable human-activity monitoringand personal healthcare. Adv. Mater. **28**(22), 4338–4372 (2016)
48. Vergara, M., Agost, M.J., Gracia-Ibáñez, V.: Dorsal and palmar aspect dimensions of hand anthropometry for designing hand tools and protections. Hum. Factors Ergon. Manuf. Serv. Ind. **28**(1), 17–28 (2018)
49. Verrillo, R.T., Bolanowski, S.J., Jr.: The effects of skin temperature on the psychophysical responses to vibration on glabrous and hairy skin. J. Acoust. Soc. Am. **80**(2), 528–532 (1986)
50. Visell, Y.: Tactile sensory substitution: models for enaction in HCI. Interact. Comput. **21**(1–2), 38–53 (2008)
51. Wang, S., Xu, J., Wang, W., Wang, G.J.N., Rastak, R., Molina-Lopez, F., Chung, J.W., Niu, S., Feig, V.R., Lopez, J., et al.: Skin electronics from scalable fabrication of an intrinsically stretchable transistor array. Nature **555**(7694), 83 (2018)
52. Xu, X.S., Welcome, D.E., McDowell, T.W., Wu, J.Z., Wimer, B., Warren, C., Dong, R.G.: The vibration transmissibility and driving-point biodynamic response of the hand exposed to vibration normal to the palm. Int. J. Ind. Ergon. **41**(5), 418–427 (2011)
53. Yang, J., Ran, Q., Wei, D., Sun, T., Yu, L., Song, X., Pu, L., Shi, H., Du, C.: Three-dimensional conformal graphene microstructure for flexible and highly sensitive electronic skin. Nanotechnology **28**(11), 115501 (2017)
54. Yousef, H., Boukallel, M., Althoefer, K.: Tactile sensing for dexterous in-hand manipulation in robotics-a review. Sens. Actuators A: Phys. **167**(2), 171–187 (2011)
55. Zhu, Y., Li, J., Cai, H., Wu, Y., Ding, H., Pan, N., Wang, X.: Highly sensitive and skin-like pressure sensor based on asymmetric double-layered structures of reduced graphite oxide. Sens. Actuators, B Chem. **255**, 1262–1267 (2018)

Chapter 6
Spatiotemporal Haptic Effects via Control of Cutaneous Wave Propagation

Abstract A perspective informing the research in this book is that new haptic technologies should be designed to account for the mechanisms of human touch sensing, including biomechanical processes. The results from the preceding chapters provide a view of tactile sensing as being mediated via the transmission of viscoelastic waves in the skin. To illustrate how these findings can inform haptic technology engineering, this chapter presents a new method for rendering evocative haptic effects by exploiting a dominant property of this wave process: frequency-dependent damping. It uses full-field optical vibrometry to show that vibrations introduced at the fingertip elicit waves in the finger that propagate proximally toward the hand, with travel distances decreasing rapidly with frequency. Based on the results, this chapter presents a new design of haptic effects producing wave fields that expand or contract in size and can be delivered via a single actuator. In a perception experiment, subjects accurately (median >95%) identified these stimuli as expanding or contracting without prior exposure or training. These findings demonstrate how the physics of waves in the skin can be exploited for the design of spatiotemporal tactile effects that are practical and effective.

Disclaimer

Previously published as: B. Dandu*, Y. Shao*, A. Stanley, Y. Visell, Spatiotemporal Haptic Effects from a Single Actuator via Spectral Control of Cutaneous Wave Propagation. *IEEE World Haptics Conference (WHC)*, Jul 2019, 425–430; DOI: 10.1109/WHC.2019.8816149. (*contributed equally) Reproduced here by permission of IEEE.

6.1 Introduction

The skin is a distributed sensory medium whose infinitely many coupled degrees of freedom are excited in complex ways during tactile interactions. A central challenge in haptic engineering is to find methods for stimulating this continuous medium via practical devices with few mechanical degrees of freedom.

It has long been observed that locally applied vibrations evoke mechanical waves distributed in the skin [10, 14, 20]. Such processes partly explain somatosensory specializations such as the large receptive fields associated with vibration-sensitive Pacinian corpuscle (PC) afferents [13, 21]. During manual activities, touching an object also excites propagating mechanical waves in the skin that reach remote locations [7, 12, 18, 22]. These waves propagate in a manner that reflects not only the anatomy and mechanics of the skin but also the properties of touched objects, the locations of contact with the skin, and the frequency content of tactile inputs [16, 18]. To date, the influence of propagating waves on tactile perception is not fully understood. While the transmission of vibrations in the skin has been considered to affect the performance of haptic devices [20], it is rarely accounted for in their design.

Many spatiotemporal vibrotactile phenomena, including apparent motion, saltation, funneling, and contrast phenomena, have been discovered, beginning in the 1900s and extending to the present day [5, 8, 9, 11, 17, 19, 24]. Such effects often reflect the spatiotemporal integration of cutaneous vibrotactile stimuli that are (in most cases) applied at multiple skin locations. For example, in 1957, von Békésy studied the spatial integration of vibrotactile sensations elicited by stimuli applied to the forearm skin, which he considered as a model system for understanding the cochlea [3].

The present work is inspired by research on cochlear auditory processing for which von Békésy received the 1967 Nobel Prize [23]. Using methods analogous to those we employ here, he combined optical measurements of basilar membrane vibrations with mechanical modeling in order to deduce that the selective transmission of lower frequency waves to greater distances provides a tonotopic spatial mapping in the basilar membrane. This mechanical process underlies frequency encoding in early auditory processing. In the vibrotactile system, an analogous frequency-dependent transmission of vibrations in the skin has been deduced to give rise to distinct population responses in cutaneous mechanoreceptors [12], but the implications for tactile perception are not fully understood.

Here, we present a new method for rendering unique, spatiotemporal vibrotactile stimuli via a single actuator. Our method exploits the viscoelastic properties of the skin, which cause propagating waves to decay in a frequency-dependent manner with increasing distance from the location at which they are applied. We hypothesized that by designing the frequency content of locally applied vibrotactile signals, we could control the spatial extent of propagating waves they excite in the skin. We further hypothesized that this could be exploited to enable a single actuator to generate spatiotemporal haptic effects.

To investigate this, we first considered a physical description for the frequency-dependent transmission of vibrotactile signals in the skin. We empirically assessed the transmission of such signals using in vivo measurements captured via optical vibrometry. These results confirmed that the frequency content of locally applied vibrotactile stimuli determines their spatial extent in the skin, with low frequencies reaching greater distances, as in the mammalian cochlea. To demonstrate the utility of these results for haptic engineering, we design vibrotactile stimuli that excite spatially expanding or contracting fields of vibration in the skin. In a perception experiment, we demonstrate that subjects perceive these stimuli as expanding or contracting, without prior exposure or training.

In the following sections, we describe the mechanics of viscoelastic waves in the skin, yielding predictions that we test in in vivo optical vibrometry measurements. We then present the design of new methods for synthesizing spatiotemporal vibrotactile stimuli from a single actuator, informed by the vibrometry results. We also present a behavioral experiment in which we demonstrate that the spatiotemporal properties of these designed stimuli are reflected in how they are perceived. We conclude with a discussion of the results and their implications for haptic device engineering and for the rendering of haptic effects via the control of waves in the skin.

6.2   Viscoelastic Waves in the Skin

Tactile sensation arises from mechanical stresses and strains felt via numerous cutaneous afferents. An idealized physical model of the transmission of such strains is an elastic wave equation

$$\rho \frac{\partial^2}{\partial t^2} \boldsymbol{\xi}(\mathbf{x}, t) = ((K + \mu/3)\nabla)\nabla \cdot \boldsymbol{\xi}(\mathbf{x}, t) + \mu \nabla^2 \boldsymbol{\xi}(\mathbf{x}, t), \qquad (6.1)$$

where $\boldsymbol{\xi}(\mathbf{x}, t)$ is the time-varying strain vector, $\mathbf{x}$ is position, t is time, ρ is density, and K and μ are the bulk and shear moduli [4], respectively. Solutions may be written as expansions in harmonic plane waves, $\boldsymbol{\xi}(\mathbf{x}, t) = e^{j(\mathbf{k} \cdot \mathbf{x} - \omega t)}$, where $\omega = 2\pi f$ is the angular frequency, $\mathbf{k}$ is the wave vector, and the wave velocity $\mathbf{v} = (\omega/k)\hat{\mathbf{k}}$. At vibrotactile frequencies, mechanical transmission in bulk occurs primarily via transverse (shear) waves, $\boldsymbol{\xi}_T(\mathbf{x}, t)$ satisfying $\mathbf{k} \cdot \boldsymbol{\xi}_T = 0$. Such waves appear to travel in soft tissues rather than via bone [14, 18], but the relative contribution of tissue types (including dermis, tendon, and muscle) is unclear. Shear waves travel at speeds $c_T = \sqrt{\mu/\rho}$. For soft tissues, the speeds, $c_T < 30$ m/s, are much lower than those of compression waves ($c > 1500$ m/s) [2, 14]. Near the skin surface, surface waves (including Rayleigh waves) develop, with speeds similar to those of bulk shear waves.

Soft tissues are viscoelastic. Waves in such tissues are damped and dispersive, yielding frequency-dependent damping $\delta(f)$ and speeds $c(f)$ [1, 12, 15]. In a linear viscoelastic model, damping imparts complex wavenumbers, $k = k_1 + i\delta$, to harmonic plane wave solutions, where δ is the damping factor, and $k_1 = 2\pi f/c(f)$.

In biological tissues, damping increases approximately linearly with frequency, $\delta = \alpha f$. Harmonic components thus satisfy

$$\boldsymbol{\xi}(x, t) = e^{-\alpha x f} e^{j2\pi f(x/c(f) - t)}. \qquad (6.2)$$

This describes an oscillating wave that decays exponentially with distance. It causes higher-frequency waves to attenuate over shorter distances than lower-frequency waves. An arbitrary plane-polarized wave has a Fourier expansion

$$\xi(x, t) = \int_{-\infty}^{\infty} df \; \phi(f) e^{-\alpha x f} e^{j2\pi f(x/c(f)-t)}. \tag{6.3}$$

Here, $\phi(f)$ is the amplitude and phase of a frequency component f, which decays over length scale $x_D = (\alpha f)^{-1}$. Thus, the relative weighting in $\phi(f)$ of low- and high-frequency content determines the distance that a propagating wave is expected to travel before attenuating. This suggests that the spatial extent of vibrotactile stimuli is greatly affected by their frequency content. We test this prediction using vibrometry experiments and use it to guide the design of spatiotemporal haptic effects.

6.3   In Vivo Vibrometry Experiments

We assessed the frequency dependence of the spatial propagation of vibrations in the hand via time-resolved optical vibrometry. This provided noncontact measurements of skin vibrations at numerous points on the glabrous hand surface in response to vibration inputs applied to the distal end of the finger. We analyzed this data to relate the frequency content of the stimuli to the spatial distribution of skin vibrations they excited in the hand.

Subjects

A total of seven subjects volunteered for the experiment (5 male; 20–45 years of age). All gave their informed, written consent. The experiment was approved by the Human Subjects Committee of the University of California, Santa Barbara.

Apparatus

Cutaneous vibrations were captured via noncontact scanning laser Doppler vibrometer (SLDV; model PSV-500, Polytec, Inc., Irvine, CA). The right hand of each participant was positioned within the SLDV field of view. The hand was stabilized in an open posture via custom 3D printed brackets affixed to five fingernails via adhesive tape (Fig. 6.1). The arm, hand, and brackets were supported via a pneumatically isolated table. Subjects were seated in a reclined chair raised to a height at which their arm could remain relaxed.

Vibration stimuli were applied normal to the tip of digit 2 along the axis of the finger via an electrodynamic actuator (Type 4810, Brüel & Kjær, Denmark) driven by a laboratory amplifier (PA-138, Labworks Inc.). Prior to the experiment, we measured the frequency response of the amplifier–actuator system to be flat over the entire measurement range (± 3 dB). The finger was attached to the actuator along

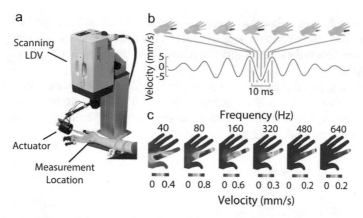

Fig. 6.1 a Measurement apparatus. Hands were positioned within the field of view of the SLDV, which captured the velocity of skin vibrations across the entire volar hand surface in response to stimuli applied at the tip of digit 2. **b** The stimuli elicited propagating waves in the skin. The data comprises one trial for one participant, with the 80 Hz windowed stimulus. **c** The RMS velocity of skin oscillations illustrate the frequency-dependent spatial extent of the wave fields elicited in the skin (shown here for one subject). Low frequencies stimuli excited waves that extended farther in the finger

a $49\,\mathrm{mm}^2$ interface via adhesive tape to prevent the skin from decoupling during actuator retraction.

Stimuli

Two types of vibration stimuli were applied via the actuator: periodic frequency sweep (chirp) and windowed sinusoidal signals. The frequency sweep extended 20–1000 Hz. The driving signals were generated by the PSV-500 system, ensuring sample-accurate synchronization of the driving signal with the measurement. The sinusoidal frequencies ranged from 40 to 640 Hz in steps 40 Hz. To avoid transient artifacts, we applied a Hanning window function to each sinusoid, with a duration of 10 cycles (thus, longer for low-frequency signals). To avoid interference between different stimuli, a pause of 150 ms was enforced between measurements. Amplitudes were selected to provide sufficient signal-to-noise ratio and to be comfortable even during extended stimulation.

Procedure

After each participant was seated at the apparatus, a 3D scan of the volar hand surface was performed via the integrated geometry scan unit of the SLDV. Vibrometry measured the velocity of skin motion normal to the volar hand surface at 300 points that were equally distributed across the surface. We selected this spatial resolution because we estimated it to be approximately 10× finer than the wavelength of cutaneous vibrations in the tested frequency range. The measurement sampling frequency was 20 kHz.

For each scanned measurement point, the finger was presented with all stimuli while data was collected at the measurement point. The vibrometer captured the vibration velocity in a direction normal to the volar hand surface. The experiment took a total of 80 min per subject.

Analysis

The data consisted of velocity signals describing skin vibrations at 300 measured locations for each of the 16 signals and 7 subjects. We focused our analysis on the skin vibrations elicited by the chirp signals. We reconstructed the time domain response at different frequencies and computed root-mean-squared (RMS) averages for each measurement trial, frequency, and participant. This yielded RMS velocities $v(\mathbf{x})$ at the 3D positions $\mathbf{x} = (x_1, x_2, x_3)$ of all measurement locations. To assess the spatial extent of skin vibrations elicited for different stimulus frequencies, we evaluated the RMS velocities v along a straight line extending from the actuator probe to the base of the hand. For each frequency, velocity values at 200 points on this line were computed via interpolation. For each frequency f, we computed the median spatial extent of skin vibrations as the distance $D(f)$ up to which the signal energy reached half of the total, averaged across all subjects. We fit $D(f)$ to a power law, $D(f) = c_1 f^{-c_2}$, where c_1 and c_2 were the fitting parameters.

Results

The results indicate that stimulation at the fingertip elicited time-dependent wave fields that reflected the stimulation frequency (Fig. 6.1a) and extended proximally from the fingertip (Fig. 6.1b), reaching the palmar surface within 10 ms.

Consistent with predictions from wave mechanics, the vibration-elicited waves decayed with the distance they traveled from the stimulation point, and did so in a frequency-dependent manner (Fig. 6.2). For all subjects, the highest frequency stimuli ($f = 640$ Hz) elicited cutaneous waves of greatest amplitude (velocity, m/s) near the fingertip, where they were applied and decayed rapidly over the course of 40 mm. Near the base of the finger, 80 mm distant, these vibration velocities were attenuated by an average of 95%. In contrast, the lowest frequency stimuli (40 Hz) elicited cutaneous waves that increased in amplitude, reaching maxima near the base of the finger. We attributed this increase to two factors. First, to an effect of spatial oscillation. The model predicts such a wave to have the form $\exp(-\alpha f) \exp(j(kx - \omega t))$, where $k = 2\pi/\lambda$, with λ being the wavelength. At low frequencies, λ is large, yield-

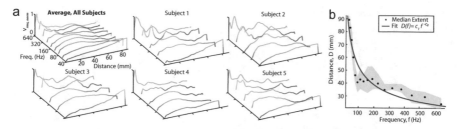

Fig. 6.2 The amplitude (normalized rms velocity $v(f)/\max_f v(f)$) of propagating waves decreased with distance in a frequency-dependent manner. **a** The magnitude of skin responses varied with distance from the location at which they were applied on the fingertip and the frequency content of the stimulus. Distances extend proximally along the midline of the finger. **b** The median spatial extent of vibrations decreased supralinearly with increasing frequency. This reflected the spatial decay and spatial wavelength of the propagating wave. A power law fit $D(f) = 435.5 f^{-0.46}$ provided a good fit ($R^2 = 0.74$) to the measurements. The shaded region represents the quartiles of distribution of the measurements

ing a gradual increase superimposed on the gradual decay with distance. Second, the stimuli were applied parallel to the finger axis. In preliminary testing with a three-axis vibrometer, we found the conversion of axial stimuli into vibrations normal to the volar hand surface to develop over the course of a fraction of a wavelength. Thus, the one-axis measurements may not capture all energy in the fingertip at low frequencies. As frequency increased 40–640 Hz, the spatial extent $D(f)$ of cutaneous waves decreased from more than 85 mm to less than 30 mm (Fig. 6.2b). This was adequately captured by a power law fit ($R^2 = 0.74$). The deviation from exponential decay could be attributed to the aforementioned spatial oscillation (consistent with the model), and to mode conversion, as noted above. Such a wave pattern is also similar to a standing wave mode. Results were similar for all participants. We attributed inter-subject differences to variations in hand geometry and contact conditions. Across subjects, the standard deviation of the normalized amplitude $v(f)/\max_f v(f)$ was 0.03, underlining the consistency of this effect.

6.4 Design of Spatiotemporal Haptic Effects

The in vivo measurements were consistent with theoretical predictions and reflected that low-frequency ($f < 160$ Hz) stimuli-excited wave extended throughout the finger, whereas damping confined waves were excited by high frequencies ($f > 160$ Hz) near to their locus of application (Fig. 6.1).

The vibration measurement procedure entailed the successive application of N brief sinusoidal stimuli with frequencies f_k that increased monotonically, $f_{k+1} > f_k, k = 1, 2, \ldots, N - 1$. This yielded skin vibrations that successively decreased in spatial extent. Participants reported that these increasing frequency sequences elicited the sensation that the spatial extent of vibrations was contracting over time. In contrast, when we tested sinusoidal stimuli with constant frequency, the spatial

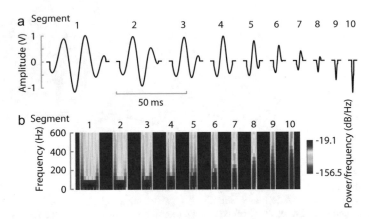

Fig. 6.3 The tactile effects were designed to elicit sensations of spatial expansion or contraction via sequences of segments with respectively increasing or decreasing frequency content. For contracting effects, the segments were wavelet-like signals **a** of decreasing duration, and **b** logarithmically increasing bandwidth

extent of skin vibrations was difficult to discern without special attention. (This might explain why the effects we describe have not received greater attention.)

Informed by these findings, we designed vibrotactile effects comprising sequences of N short segments $x_k(t)$, $k = 1, 2, \ldots, N$ of successively increasing or decreasing frequency content. We designed the changes in frequency to span a range from 12.5 333 Hz, which matched the range over which the spatial extent of skin vibrations changed most rapidly (Fig. 6.2b). We first tried sequences of sinusoidal segments. However, we observed them to elicit ancillary sensations of ascending or descending "tones" that were distracting. To avoid this, we instead designed sequences of wavelet-like segments $x_k(t)$, whose energies were more broadly distributed in frequency (Fig. 6.3). The segments consisted of scaled window functions modulated by a fixed frequency, $f_0 = 25\,\text{Hz}$.

$$x(t) = \sum_{k=1}^{N} x_k(t - r_k), \tag{6.4}$$

$$= \sum_{k=1}^{N} A_k\, g_k(t - r_k)\, \sin(2\pi f_0(t - r_k)). \tag{6.5}$$

Here, $T = 100$ ms is a short delay time between segments, and A_k is a gain factor that compensated for differences in the perceived magnitude of the segment signals. The onset time of segment k is $r_k = \sum_{j=0}^{k-1}(\tau_j + T)$. The window function $g_k(t)$ is a truncated Gaussian of duration τ_k and standard deviation $\sigma = \tau_k/5$

$$g_k(t) = \text{rect}\,(t/\tau_k)\,\exp\left(\frac{-t^2}{2\sigma^2}\right), \quad \sigma = \tau_k/5 \qquad (6.6)$$

$$G_k(f) = \sqrt{2\pi}\sigma\,\text{sinc}(f\tau_k) \star e^{-2(\pi f \sigma)^2}, \qquad (6.7)$$

where $\star$ denotes convolution in the frequency domain and $G_k(f) = \mathcal{F}(g_k(t))$ is the window spectrum. The spectrum, $X_k(f) = \mathcal{F}(x_k(t))$, of each signal segment is given by

$$X_k(f) = A_k G_k(f) \star \frac{j}{2}[\delta(f + f_0) - \delta(f - f_0)] \qquad (6.8)$$

It is centered at frequency f_0 with a segment bandwidth of approximately $f_k = 1/\tau_k$. For the concentrating stimulus, we designed signals $x(t)$ with N segments having frequency bandwidths f_k that increased in a logarithmic series over time, from $f_1 = 12.5\,\text{Hz}$ to $f_N = 333\,\text{Hz}$. For the expanding stimulus, the sequence was reversed. Stimuli used in the experiments had $N = 10$ segments.

6.5 Perception of Spatiotemporal Haptic Effects

We next sought to determine whether our theoretical and mechanical findings linking the frequency content of tactile signals to their spatial extent would be reflected in perception. To this end, we designed a simple experiment in which participant classified either the wave-based effects (described in the preceding section) or control signals as "expanding" or "contracting".

Participants

Fifteen participants volunteered for the experiment (12 male; 19–30 years of age). All subjects were naïve to the purpose of the experiment and gave their written informed consent. The experiment was conducted according to the protocol approved by the UCSB Human Subjects Committee.

Apparatus

Vibration stimuli were applied to the distal end of the index finger using hardware identical to that used in the measurements. Subjects were seated, with their index finger in contact with a flat surface (diameter 15 mm) on the actuator (Fig. 6.4c). The finger was held at a 45° angle and a contact force of 2 N (a force gauge was used for comparison). Subjects wore foam earplugs (attenuation rating 33 dB) and

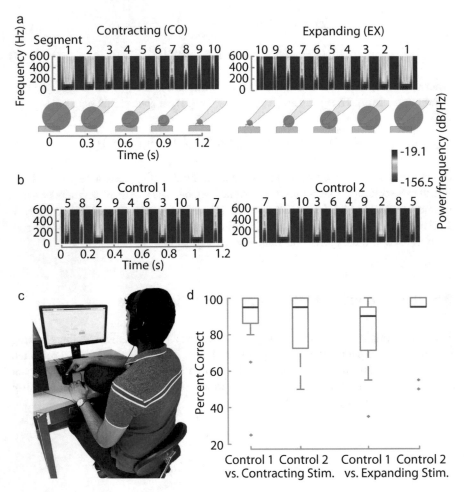

Fig. 6.4 Perceptual experiment. **a** Spectrogram of the designed stimuli that produce contracting (CO) and expanding (EX) sensations and associated animations used. **b** Spectrogram of the control stimuli, with permuted segments. **c** Experiment setup. Subjects contacted the actuator with their right index finger. **d** Results of the experiment for four pairs of designed vs. control stimuli. Subjects consistently identified the designed stimuli

circumaural headphones playing pink noise sufficiently loud to mask feedback from the actuator. The experiment was automated via a computer's graphical user interface.

Stimuli

The stimuli consisted of the designed "expanding" (EX) or "contracting" (CO) stimuli described in the preceding section (Fig. 6.4a). We also designed two different control stimuli (Fig. 6.4b). The control stimuli were composed of sequences of the same signal segments as in the EX and CO stimuli, in permuted (shuffled) order. Because we hypothesized that the sequence of changes in frequency extent of the EX or CO stimuli would cause them to be perceived as expanding or contracting, we selected permutations so that there was no systematic increase or decrease in frequency extent over the course of each entire control signal, or over the first or second half separately (Fig. 6.4b).

Procedure

The experiment was based on a two alternative forced choice task in which subjects were presented with one designed stimulus (EX or CO) and one control stimulus. Subjects were also presented with an animated visual representation of "expansion" or "contraction" at the fingertip matching the presented designed stimulus (Fig. 6.4a). We did not describe what these represented, nor did we present the stimuli, which subjects felt for the first time on the first recorded trial of the experiment. In pretesting, we determined that visual representations were understood better than text labels. Subjects were allowed to experience the two stimuli and asked to report the stimuli which caused a perceptual sensation closest to what is indicated in the displayed animation. The stimuli were presented in random order and their position was randomized on the computer screen. We randomly selected one designed and one randomly control stimulus for each trial, yielding four possible combinations. Each of the designed or control stimulus was presented a total of 40 times, for a total of 80 trials per participant. In a post-experiment questionnaire subjects were asked to describe the strategy they used to respond. They then felt the designed stimuli again and were asked to describe them.

Analysis

We computed the percentage correct as the median proportion of responses that assigned the designed stimuli to the corresponding visual depiction for each of the four combinations for each subject. The proportions were computed from all presentations (including either control stimulus). We considered a percentage correctness of 75% as a threshold for a positive effect relative to chance performance (50%). Because the distributions violated normality assumptions, we used nonparametric one-sample Wilcoxon signed-rank tests. We tested the null hypothesis that the median was less than or equal to 75%. We also assessed the effect of the control stimulus choice using Mann-Whitney U-test.

Results

Without training or prior exposure, subjects consistently associated the designed stimuli with the visual representations of "expanding" or "contracting". The designed stimuli were selected with median probability 96.7% (CO stimuli: 97.5%, EX: 95%). The control stimuli were selected with median probability 3.3%. These values were significantly greater than a positive effect threshold of 75% ($p = 0.0163$ and $p = 0.0039$ for CO and EX). There was no significant effect of the control stimulus presented ($p > 0.7$ and $p > 0.08$ respectively).

Discussion

The results indicate that the designed stimuli were consistently associated with the corresponding visual depiction. This suggests that subjects felt them as respectively expanding or contracting, consistent with predictions from theory and mechanical observations, without any prior exposure or training. The high median accuracy reflects the robustness of these associations. The associations were unaffected by the control stimulus used for the comparison.

After repeated exposure, participants would have had little difficulty in identifying either the designed or control stimuli, due to the differences in timing and frequency content. Consequently, they could have based their decisions on arbitrary cognitive criteria. However, a priori, we would expect such criteria to result in a bistable response pattern. In our experiment, no subjects consistently inverted the aforementioned associations to expansion and contraction. Further, the control stimuli were selected with a very low median response rate (3.3%). Together, these results indicate that the effect was not bistable, and suggest that arbitrary criteria were not applied.

We conclude that the increasing or decreasing frequency content of the designed stimuli caused them to be perceived as spatially expanding or contracting. This conclusion is supported by the questionnaires, in which participants spontaneously described sensations such as "concentrating to/spreading from the fingertip" and proceeding "from bigger portion of the finger to smaller portion/vice versa."

The designed stimuli were selected with greater frequency after the first few trials, from 87.5% (median percentage for trials 1–4) to 100% (trials 5–20). Because participants did not feel any stimuli prior to trial 1, this could reflect increasing familiarity with the protocol. It could also reflect strengthened associations after repeated exposure.

6.6 Conclusions

A central challenge in the engineering of tactile interfaces is to stimulate the continuous medium of the skin via practical mechanical instruments with very few mechanical degrees of freedom. In this work, we show how to use a single actuator to generate tactile stimuli with dynamically controlled spatial extent. The methods are based on the physics of frequency-dependent damping of propagating waves in the skin. We provided empirical evidence for this through new observations from full-field optical vibrometry. This revealed that the frequency content of locally applied vibrotactile stimuli shapes their spatial extent in the skin. We demonstrated the utility of these results by designing expanding and contracting vibrotactile stimuli that can be delivered by a single actuator. Using a perception experiment, we demonstrated that the designed stimuli were perceived as respectively expanding or contracting, consistent with predictions from theory and mechanical observations. These findings demonstrate how the wave physics of the skin can be exploited to design methods for spatiotemporal haptic feedback that are practical and perceptually evocative.

We showed that these phenomena can be exploited to realize spatial effects using a single actuator. It is interesting to consider how they may be integrated in multi-actuator configurations to generate effects of heightened tactile apparent motion. We aim to explore this in future work.

Although the results are promising, several issues merit further investigation. First, the wave phenomena involved are complex, and further research is needed to fully characterize them. Second, the experiments involved a simpler binary forced choice paradigm and further research is needed in order to clarify how such effects are perceived. Further, this study focused primarily on the fingertip, the region of the body with which we frequently contact objects. However, the generality of these physical phenomena suggests that similar effects could be elicited at other body locations.

References

1. Achenbach, J.: Wave Propagation in Elastic Solids. Elsevier (2012)
2. Azhari, H.: Basics of Biomedical Ultrasound for Engineers. Wiley (2010)
3. Békésy, G.: Sensations on the skin similar to directional hearing, beats, and harmonics of the ear. J. Acoust. Soc. Am. **29**(4), 489–501 (1957)
4. Blandford, R., Thorne, K.: Ph 136: Applications of Classical Physics. California Institute of Technology, Pasadena (2003)
5. Burtt, H.E.: Tactual illusions of movement. J. Exp. Psychol. **2**(5), 371 (1917)
6. Dandu, B., Shao, Y., Stanley, A., Visell, Y.: Spatiotemporal haptic effects from a single actuator via spectral control of cutaneous wave propagation. In: 2019 IEEE World Haptics Conference (WHC), pp. 425–430. IEEE (2019)
7. Delhaye, B., Hayward, V., Lefèvre, P., Thonnard, J.L.: Texture-induced vibrations in the forearm during tactile exploration. Front. Behav. Neurosci. **6**, 37 (2012)
8. Geldard, F.A., Sherrick, C.E.: The cutaneous "rabbit": a perceptual illusion. Science **178**(4057), 178–179 (1972)
9. Israr, A., Poupyrev, I.: Tactile brush: drawing on skin with a tactile grid display. In: Proceedings of the SIGCHI Conference on Human Factors in Computing Systems, pp. 2019–2028 (2011)
10. Johansson, R.S., Vallbo, Å.B.: Tactile sensory coding in the glabrous skin of the human hand. Trends Neurosci. **6**, 27–32 (1983)
11. Kirman, J.H.: Tactile apparent movement: The effects of interstimulus onset interval and stimulus duration. Percept. Psychophys. **15**(1), 1–6 (1974)

12. Manfredi, L.R., Baker, A.T., Elias, D.O., Dammann III, J.F., Zielinski, M.C., Polashock, V.S., Bensmaia, S.J.: The effect of surface wave propagation on neural responses to vibration in primate glabrous skin. PloS One **7**(2) (2012)
13. Manfredi, L.R., Saal, H.P., Brown, K.J., Zielinski, M.C., Dammann, J.F., III., Polashock, V.S., Bensmaia, S.J.: Natural scenes in tactile texture. J. Neurophysiol. **111**(9), 1792–1802 (2014)
14. Moore, T.J.: A survey of the mechanical characteristics of skin and tissue in response to vibratory stimulation. IEEE Trans. Man-Mach. Syst. **11**(1), 79–84 (1970)
15. Pereira, J., Mansour, J., Davis, B.: Dynamic measurement of the viscoelastic properties of skin. J. Biomech. **24**(2), 157–162 (1991)
16. Schäfer, H., Wells, Z., Shao, Y., Visell, Y.: Transfer properties of touch elicited waves: effect of posture and contact conditions. In: World Haptics Conference (WHC), 2017 IEEE, pp. 546–551. IEEE (2017)
17. Schneider, O.S., Israr, A., MacLean, K.E.: Tactile animation by direct manipulation of grid displays. In: Proceedings of the 28th Annual ACM Symposium on User Interface Software & Technology, pp. 21–30. ACM (2015)
18. Shao, Y., Hayward, V., Visell, Y.: Spatial patterns of cutaneous vibration during whole-hand haptic interactions. Proc. Natl. Acad. Sci. **113**(15), 4188–4193 (2016)
19. Sherrick, C.E., Rogers, R.: Apparent haptic movement. Percept. Psychophys. **1**(3), 175–180 (1966)
20. Sofia, K.O., Jones, L.: Mechanical and psychophysical studies of surface wave propagation during vibrotactile stimulation. IEEE Trans. Haptics **6**(3), 320–329 (2013)
21. Vallbo, A.B., Johansson, R.S., et al.: Properties of cutaneous mechanoreceptors in the human hand related to touch sensation. Hum. Neurobiol. **3**(1), 3–14 (1984)
22. Visell, Y., Shao, Y.: Learning constituent parts of touch stimuli from whole hand vibrations. In: 2016 IEEE Haptics Symposium (HAPTICS), pp. 253–258. IEEE (2016)
23. Von Békésy, G., Wever, E.G.: Experiments in hearing, vol. 8. McGraw-Hill, New York (1960)
24. Whitchurch, A.K.: The illusory perception of movement on the skin. Am. J. Psychol. **32**(4), 472–489 (1921)

Chapter 7
Conformable Distributed Haptic Feedback to Large Areas of the Skin

Abstract An overarching goal of this book is to design new methods of haptic feedback informed by the properties of the human haptic system. As the previous chapters illustrate, the soft mechanics of the skin play an important role in touch sensing. In contrast, existing tactile feedback devices frequently rely on rigid actuated elements. This chapter presents a very different approach, in the form of a soft and conformable tactile display that can provide distributed dynamic haptic feedback to large areas of the skin. This display combines electrostatic attraction with hydraulic amplification provided by a liquid dielectric encapsulated in a compliant pouch. Voltage supplied to six pairs of opposed hydrogel electrodes generates dynamic variations in pressure on the encapsulated liquid. Mechanical amplification by the liquid enables the device to render tactile feedback with substantial displacements (>2 mm) and forces (>0.8 N) via a thin (<3.5 mm) compliant surface with a large active area (75 cm^2). This chapter presents the design of the device from the perspective of its performance, reliability, and safety. It describes a fabrication method that enables the device to be easily reproduced by others. The user study shows that the device can produce unique haptic experiences, such as fluid-mediated haptic effects of motion across the skin.

Disclaimer

Previously published as: Y. Shao, S. Ma, SH. Yoon, Y. Visell, J. Holbery, SurfaceFlow: Large Area Haptic Display via Compliant Liquid Dielectric Actuators. *IEEE Haptics Symposium (HAPTICS)*, Mar 2020, 815–820; DOI: 10.1109/HAPTICS45997.2020.ras.HAP20.23.0f334629. Reproduced here by permission of IEEE.

7.1 Introduction

Among many haptic experiences in daily life, only a few are well reproduced using existing devices [3]. One challenge is to reproduce the sensation of touching a soft, deformable object with large skin contact. A device that can provide such feedback may be useful in many applications. For example, in medical training, a haptic simulator can train physicians and improve their skill of palpation [21]. Another application is social virtual reality, allowing interpersonal touch that enhances social presence in a virtual environment [11].

It is challenging to render such experiences with conventional haptic feedback devices, many of which involve rigid elements whose attributes impair the ability of such devices to simulate many naturally soft and organic materials [4, 8, 18]. Recently, several skin-conformable haptic display devices based on pneumatic or hydraulic sources have been developed [17, 19, 22, 24, 25]. However, these often entail pumps that are bulky and slow to respond, making them hard to control [2]. Many haptic wearables, including gloves, integrate vibration motors. The sensations produced by such devices are rarely similar to what is felt during natural touch. There are several reasons for this, including their bandwidth limitations, which prevent them from producing controlled low-frequency haptic feedback [15, 23]. While such shortcomings can be addressed via exoskeleton designs [14], such mechanical structures often further reduce the naturalness.

Material technologies for haptic feedback are rapidly advancing [2], including electrostatic actuator technologies [12]. Such actuators comprise a compliant dielectric layer sandwiched between two conductive layers. A voltage applied to the conductors produces an electric field generating charges with opposite polarities at two sides of the dielectric. Electrostatic forces act to reduce the distance between the conductive layers. In dielectric elastomer actuator (DEA), the compliance of a dielectric makes it possible for a voltage in the 10 kV range to generate small, dynamic displacements. However, such actuators have rarely been used in haptics applications, due to several commonly occurring shortcomings, including restricted dynamic range, bandwidth, and fragility. A recently proposed variation on DEAs is the hydraulically amplified electrostatic actuator [1]. These devices employ a liquid dielectric, rather than a solid, in combination with an insulating thin film layer. The liquid undergoes large bulk displacement, yielding mechanical amplification of displacement and larger forces. These devices can be driven with kV range voltages and be operated from batteries using suitable voltage converters [7, 10].

Informed by these developments, we designed SurfaceFlow, a large area haptic display based on compliant, liquid dielectric actuators. This display combines electrostatic attraction with hydraulic amplification provided by a liquid dielectric encapsulated in a compliant pouch. A voltage supplied to opposed electrodes generates dynamic variations in pressure on the encapsulated liquid. The device is lightweight and thin, and could lead to skin-like body-worn interfaces. The large area of contact improves mechanical transmission to the skin, and the soft rubber surface of the device lends it a comfortable feel, like skin-to-skin contact. The multiple electrostatic actuators integrated in the device enable it to provide both static and dynamic spatial patterns of tactile feedback to large areas of the skin. Unlike haptic feedback devices driven by electromagetic acutators, this device can operate without producing audible noise.

The rest of the chapter is organized as follows. First, we present the design and operating principles of the SurfaceFlow display device. The design is informed by considerations of performance, reliability, and safety. We next describe how to make them using a process designed to make them easy to reproduce. We then discuss the electronic control system. We present two experiments evaluating the performance of the device, one characterizing the static and dynamic mechanical performance for both force and displacement, another assessing its effectiveness in producing tactile patterns on the hand. We conclude with a discussion of the main findings and opportunities for future rescarch.

7.2 Design and Operating Mechanism

Our device was designed to contain a liquid dielectric layer, in a thin, compliant, palm-sized pouch with six pairs of opposed electrodes. Applying a voltage to select electrode pair squeezes the liquid in the gap between them, which displaces the liquid, resulting in bulk deformation of the surface (Fig. 7.1). The dielectric liquid is sealed inside an insulating pouch. Pairs of opposed hydrogel electrodes are affixed to the exterior of the pouch. For each pair of electrodes, one is grounded and positioned on the side of the device facing the skin, while positive voltage is applied to the other, which is attached to the side facing away from the skin. This structure is enclosed in a silicone protective pouch, providing further protective insulation.

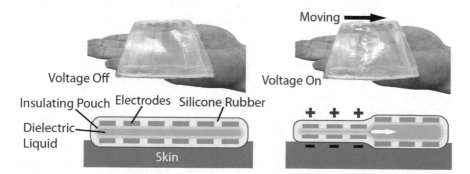

Fig. 7.1 Top row: Photos of the display device during operation on the hand. Bottom row: Simplified structure (cross-sectional view) and operating mechanism. The dielectric liquid is encapsulated by an insulating pouch, with opposed pairs of electrodes on each side. Voltages applied across pairs of electrodes generate electrostatic pressures that close the gap between them, displacing the dielectric liquid to volumes between inactivated electrodes, and producing a tactile motion on the skin

When voltage is applied across an electrode pair, an electrostatic force is produced, compressing the dielectric liquid and causing it to be displaced to volumes between other inactivated electrodes. This causes selected regions of the device to expand, yielding force and displacement feedback that are felt by the skin. The device can be restored to its original state by applying voltage to electrodes in the expanded area.

The electrostatic force produced by each pair of electrodes depends on the applied voltage, the resulting charge between them, and the distance between electrodes [20]. Our device integrates two different dielectric materials into two layers between the electrodes. Let z_i be the total thickness of the insulating pouch layer and z_f the thickness of the dielectric liquid. Since the insulating layer undergoes little deformation, z_i is nearly a constant during actuation (for essentially any value $z_f > 0$). The electrostatic force between an opposed electrode pair is given by

$$F = -\frac{dW}{d(z_i + z_f)} = -\frac{dW}{dz_f},\tag{7.1}$$

where W is the work done by closing the gap between the electrodes. The latter is given by:

$$W = \frac{1}{2}CU^2,\tag{7.2}$$

where U is the applied voltage. C is the total capacitance of the dielectric structure composed of the insulating layer and the dielectric liquid, with capacitances C_i and C_f, respectively. Since they are connected in series:

$$\frac{1}{C} = \frac{1}{C_i} + \frac{1}{C_f},$$
$$C_i = \frac{\epsilon_0\epsilon_i A}{z_i}, C_f = \frac{\epsilon_0\epsilon_f A}{z_f}.\tag{7.3}$$

Here, $\epsilon_0 = 8.854 \times 10^{-12}$ Farad/m is the free-space permittivity. ϵ_i and ϵ_f are the relative permittivities of the insulating layer and the dielectric liquid, respectively. A is the area of the electrodes. Combining equations (7.1), (7.2), and (7.3), we obtain

$$F = \frac{1}{2}\epsilon_0\epsilon_f A \frac{U^2}{(z_f + \frac{\epsilon_f}{\epsilon_i}z_i)^2}.\tag{7.4}$$

To maximize the electrostatic force that squeezes the liquid, dielectric materials with high permittivity and modest thickness are needed. Moreover, these materials should possess sufficient dielectric strength to avoid electrical discharge. To achieve this, transformer oil (Envirotemp FR3, Cargill, Inc.), which is designed to have precisely these attributes, was selected as the dielectric liquid. The outer insulating layer was made from soft silicone rubber (Ecoflex 00–30, Smooth On Inc.). The electrodes were fabricated using UV-curable conductive hydrogel (JN0917-A, Polychem UV/EB

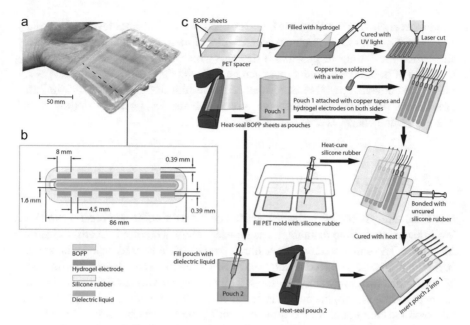

Fig. 7.2 **a** The SurfaceFlow device is soft, flexible, and translucent. **b** Cross-sectional view of the device. It has two components: An inner BOPP insulating pouch layer encapsulating the dielectric liquid and an outer insulating silicone layer. Six pairs of opposed hydrogel electrodes are positioned on opposite sides of the liquid pouch. **c** Overview of the fabrication procedure

International Corp). They are flexible and transparent. The finished device has a thickness of 3.27 mm and a weight of 34.4 g. The active display area is 75 cm^2, enabling it to provide tactile feedback to large skin areas, such as the palm (Fig. 7.2a) or forearm.

7.2.1 Design for Reliability

One challenge in using electrostatic actuators is that device failure can be caused by dielectric discharge between electrodes. Choosing an appropriate thickness, z_i, for the insulating pouch layer avoids the possibility of dielectric breakdown. However, as z_i increases, higher voltages are needed in order to produce an equivalent force (Eq. (7.4)). We selected a thin insulating material, biaxially oriented polypropylene (BOPP), with < 18 μm thickness, inspired by prior research on similar actuators [10]. Although BOPP has a dielectric breakdown strength of 700 $V/$μm, defects can reduce the dielectric strength [9]. To further improve the robustness of the device, we designed a dual-layer structure to avoid this possibility. The resulting device structure is shown in Fig. 7.2b. Two BOPP sheets with 17.5 μm thickness (MSB, Impex Global Films) are combined below each electrode, including, for each pair, an inner layer enclosing the dielectric liquid on each side and an outer layer abutting each electrode. Owing to this structure, defect-related dielectric breakdown is unlikely, since this would only be possible if defects (which are rare) occur at overlapping locations of both layers. Either layer can be replaced easily should a damage occur, making the device repairable. As discussed below, our design ensures that dielectric breakdown, which involves discharge between electrodes, poses no significant risk to a user, since charge is confined within the device.

7.2.2 Design for Safety

Electrical safety is another key consideration due to the significant operating voltages. We integrated multiple features to ensure user safety. The first is exterior electrical insulation. The entire electrostatic actuator structure is enclosed inside silicone rubber, with thickness of 390 μm and breakdown strength >13.8 V/μm. Furthermore, all electrodes are sandwiched by the BOPP sheets, providing additional electrical insulation. In addition, the device is designed to contact a user on the side in which all electrodes are connected to ground. Finally, the operating currents used by the device are low (<200 μA) and poses minimal health risks to the skin [13]. This is achieved via control electronics that limit current, as discussed in Sect. 7.3.

7.3 Fabrication

We designed a fabrication process that would enable the device to be easily reproduced. The entire fabrication procedure is shown in Fig. 7.2c. A laser cutting machine (VLS4.60, Universal Laser Systems Inc.) was used to cut materials, including BOPP and polyethylene terephthalate (PET) sheets, into desired shapes and create molds. They were cleaned with methanol and dried at room temperature. A 390 μm spacer was laser cut from PET sheet and sandwiched between two BOPP sheets. The molds were filled with 6.5 mL conductive hydrogel and cured under UV light (ECE 5000, Dymax Corporation) for 40 s. After this, electrodes were cut from the BOPP-shielded hydrogel. Two pouches were made from BOPP sheets with three sides heat-sealed. 50 μm copper tapes soldered with wire connections were attached to one pouch (pouch 1), and were then covered with hydrogel electrodes, forming electrical connections from the wires to the electrodes. Two 390 μm silicone rubber sheets were made by filling a PET mold with liquid silicone rubber and cured in an oven (Lab Companion Model OF-02, Jeio Tech Inc.) at 65 C° for 7 min. They were bonded to both sides of pouch 1 to cover the electrodes, using more liquid silicone rubber. After curing in the oven at 45 C° for 10 min, the electrodes were completely sealed below the silicone rubber layer. Finally, the second BOPP pouch (pouch 2) was filled with 12 mL dielectric liquid, then heat-sealed, and inserted into pouch 1.

7.4   Control System

A multi-channel voltage control system was developed in order to actuate each electrode pair independently (Fig. 7.3). The device is controlled by a PC through a data acquisition (DAQ) interface (Model 6255, National Instruments). An analog output from the DAQ is amplified to 8 kV via a power amplifier (Model 20/20 A-G, Trek, Inc.). The amplifier output is distributed via switched connections to six channels connected to the electrodes of the electrostatic actuators. Each switched connection is regulated by an integrated optocoupler (OC100HG, Voltage Multipliers Inc.) that is controlled using the 5 V digital output of the same DAQ. With voltage applied, the two LEDs inside the optocoupler emit light that opens the photodiode-based transistor. With the 100 Ω resistor limiting the current of the LEDs, the maximum current passing through the photodiode-based transistor is limited ($< 250\,\mu A$) in all high-voltage channels. Electrical current is also monitored by the PC in real time. The voltage is interrupted if the current exceeds the limit preset in the PC. These features further improve the reliability and safety of the device.

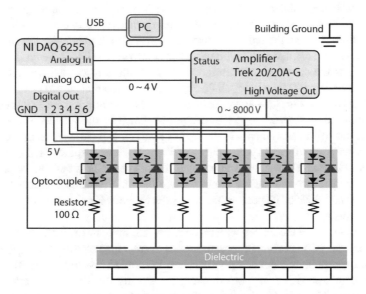

Fig. 7.3 Control system of the device. A PC-interfaced NI DAQ was used to control and monitor the status (output voltage and current) of the amplifier and independently regulate the current in each channel using optocouplers

7.5   Validation Experiments

To evaluate the performance of the SurfaceFlow display, measurements of its force output and displacement were recorded. In addition, we evaluated the devices ability to produce tactile feedback patterns on the skin, via the palm of the hand.

7.5.1   Mechanical Evaluation

A computer-controlled mechanical stimulator (Model 300C-LR, Aurora Scientific Inc.) was used to measure the force output and displacement of the device (Fig. 7.4a) during actuation. An acrylic contact plate, with the same size as the first electrode of the device, was bonded to the tip of the mechanical stimulator device. The display was placed between the contact plate and a supporting platform, with the electrode on the left overlap with the contact plate. During the measurements, electrodes 6, 5, and 4 were actuated consecutively from the right, displacing liquid to the left, causing electrode 1 (the leftmost) to move upward, pushing the contact plate against the mechanical stimulator. Evaluations were performed under two different measurement conditions. First, the unloaded displacement (UD) of the device was measured during maximum displacement with minimum preload (<0.04 N). Second, blocking force (BF) of the device was measured at the maximum force that was required to prevent displacement (threshold <0.06 mm). An example measurement in each condition is shown in Fig. 7.4b and c, respectively. Two features were extracted from the measurements in order to assess the performance of the device. The red solid line indicates the measurement obtained after averaging 100-ms measurement starting from the peak of the response curve of force (BF) or displacement (UD). This provides an estimate of the low-frequency (DC) displacements that can be produced on the skin. The two red dashed lines indicate the time instants at where the curve reached 10 and 90% of its peak value. The interval between them was the rise time of the response.

Measurements of the device with different actuation patterns were taken for both tests (UD and BF). For each, measurements were repeated five times, then both the mean and standard deviation (SD) were computed. We evaluated the impact that actuation speed has on performance, by closing the three electrode pairs at different time intervals Δt (Fig. 7.5a). The actuation voltage was set to 8 kV. In the UD condition, the maximum displacement was 2.2 ± 0.1 mm (mean $\pm$ SD) when $\Delta t = 50$ ms, and corresponding rise time was 114 ± 17 ms, as shown in Fig. 7.5b. When $\Delta t = 0$, all three pairs were closed simultaneously. However, this resulted in a relatively small displacement and longer rise time, compared to asynchronous actuation, with $\Delta t = 10, 50$ or 100 ms. The device behaved very differently in the BF condition. The force output for cases with $\Delta t \leq 100$ ms were similar. The rise time of force increased with Δt.

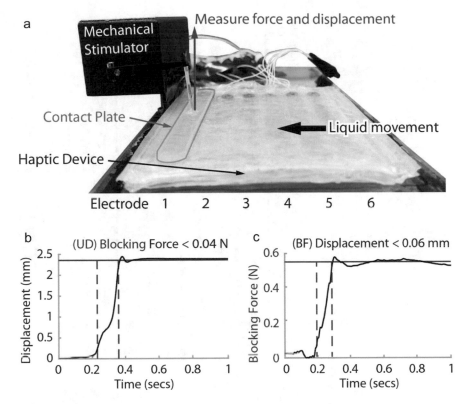

Fig. 7.4 Force and displacement measurement of the device using a mechanical stimulator. **a** The device was placed between the contact plate of the mechanical stimulator and a support platform. The contact plate overlapped the electrode on the left. Three pairs of electrodes on the right moved the dielectric liquid to the left, pushing the contact plate upwards against the mechanical stimulator. Example measurements: **b** Unloaded displacement (UD) of the contact plate when the mechanical stimulator outputs blocking force close to zero. **c** Blocking force (BF) of the contact plate when mechanical stimulator held the plate against the device without displacement. The solid red lines indicate time-averaged peak responses. The time duration between two dashed lines indicates the rising time

In addition, the performance was evaluated as a function of the actuation voltage. In this case, the device was driven by 4, 6, and 8 kV, with a fixed actuation interval $\Delta t = 50$ ms. As voltage increased, both displacement (Fig. 7.5d) and force output (Fig. 7.5e) increased and rise time decreased. There were insignificant differences between displacements at 6 and 8 kV, suggesting that, by the time 50 ms had elapsed, most liquid was displaced to the space between inactive electrodes when driven with voltage ≥ 6 kV. However, 8 kV produced a much larger force than was produced at 6 kV, as expected from the theoretical model.

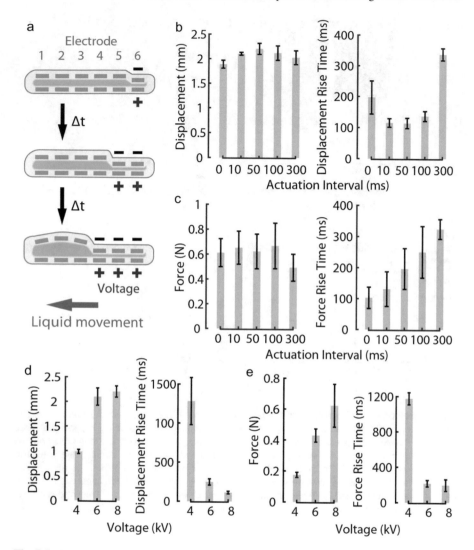

Fig. 7.5 Evaluation of effects of actuation speed and voltage using the configuration in Fig. 7.4. **a** The device was operated by closing electrode pairs consecutively with a fixed actuation interval Δt. Results: Measurements of **b** displacement (UD) and **c** force (BF) at different actuation speeds when voltage = 8 kV. Measurements of **d** displacement (UD) and **e** force (BF) with different actuation voltages when Δt = 50 ms

7.5.2   User Study

We performed a user study to assess the ability of the device to render tactile motion patterns (proximal-distal direction). Nine participants (four females and five males, aged 23–41 years old) volunteered to participate in the study. All participants are right-hand dominant. No participants reported abnormalities affecting touch perception. The study was conducted according to both institution guidelines and national regulations on human subject research. All participants gave their informed consent. The experiment setup is shown in Fig. 7.6a. The participants were seated in front of a table with the device and a PC. An interface was displayed on the PC, allowing users to control the device and submit answers. In each trial, the participants rested the right palm on the display area of the device, with electrodes 1–6 aligned from distal to proximal area of the palm (Fig. 7.6b). Through the user interface, participants played and experienced two stimuli, one at a time, and were asked to select the one that best matched the visual display (Fig. 7.6c). Participants were instructed to lift their hand after they felt a stimulus and were required to rest for 10 s, in order to allow the device to be fully restored to its original shape. Only one of the two stimuli produced tactile motion in the direction described by the visual display. The other one was either a control stimulus that changes direction (Fig. 7.6d) or a stimulus moving in the direction opposite to the visual display, yielding six combinations. In our design, only three electrodes of the device were actuated sequentially to generate a stimulus. There were two target stimuli presenting upward or downward motion (moving up by closing electrode 6, 5, 4 or down by closing 1, 2, 3) and two control stimuli (closing electrode 4, 2, 5 or 3, 5, 2). Each combination was repeated three times with actuation setting $\Delta t = 50$ ms and three times with $\Delta t = 300$ ms, as the chosen time intervals provided perceivably different tactile motion in a pilot study. Those combinations were presented in a random order, resulting in 36 trials per participant.

We logged the number of trials when participants correctly associated the stimulus they felt to the direction on the visual display. The median correct response rate was 94.4%. Mann-Whitney U tests were performed, with significance level set to $p = 0.05$. The effect of actuation speeds on identification correctness (Fig. 7.6e) was insignificant ($p = 0.61$). The effect of direction of target stimuli when discriminated against control stimuli were also evaluated (Fig. 7.6f). The identification correctness of target stimuli moving up (median 100%) and down (median 91.7%) were similar ($p = 0.37$). When differentiating the two target stimuli, seven out of nine participants had 100% correctness. The results suggest that the device can produce directional tactile motions on the skin that are easily recognizable by most users. The actuation speed and direction had little effect on the performance.

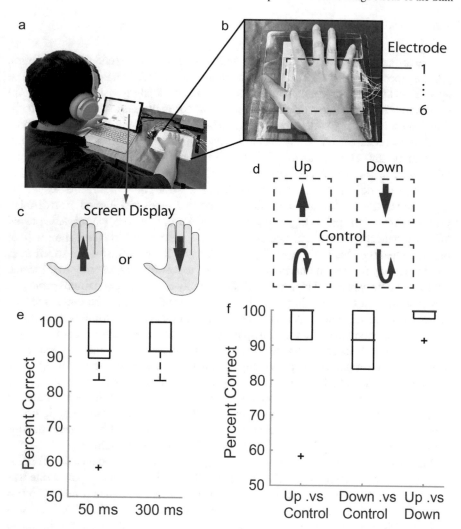

Fig. 7.6 User evaluation on tactile motion pattern rendering. **a** Experiment setup: Participants **b** rested their dominant hand on the device and **c** matched perceived motion with the figure displayed on the screen. **d** Two target (up and down) and two control stimuli were produced in the experiment. Red arrows indicate the motion of dielectric liquid. Dashed lines illustrate the boundary of the haptic display. The box plots show identification rate of target stimuli for **e** two actuation speeds and in **f** two different tasks: identifying target against control stimuli and discriminating two target stimuli

7.6 Conclusion and Future Opportunities

We developed a large area haptic display, the SurfaceFlow, based on compliant, liquid dielectric actuators. It combines electrostatic actuation with hydraulic amplification provided by a liquid dielectric encapsulated in a compliant pouch. The device can provide dynamic tactile feedback patterns to large areas of skin.

Multiple design features ensure reliable operation, repairability, and safety. We presented a fabrication method that makes the device easy to reproduce. We evaluated the performance of the device through mechanical measurements and a user study. The results demonstrate the effectiveness of the device in providing substantial and recognizable tactile feedback, including patterns of motion on the skin.

While the results are promising, several opportunities exist for extending this work. For example, while our system was operated using a lab amplifier, such a device can readily be driven by a battery and voltage converter, enabling portable use [10]. Although the voltages used can readily be provided via lab equipment or batteries, there would be advantages to enable lower operating voltages. One way to achieve this would be to employ a liquid with higher dielectric permittivity. Another would be to design a structure with multiple stacked dielectric layers. We plan to pursue both in future work. Second, the complex liquid dynamics can greatly affect the actuator outputs. Since electrostatic actuators have the merit of self-sensing [6], implementing a closed-loop control of the actuators could make the device output more controllable. Finally, the electrostatic actuators are capable of vibrating up to several hundred Hertz [10], covering most of the tactile sensitivity range of the human hand [5]. Such vibrations could enable richer haptic experiences to be rendered, such as contact events or textures. In future work, we will enhance the device to produce more complex tactile patterns and demonstrate its use cases, such as haptic rending of animate objects, inspired by comments in the post-experiment survey in which some participants reported a sensation of "heartbeat" or "petting a cat" after they had felt the tactile stimuli. We envisage applications of such displays in many domains, ranging from wearable garments for social haptics to medical training interfaces and VR displays.

References

1. Acome, E., Mitchell, S., Morrissey, T., Emmett, M., Benjamin, C., King, M., Radakovitz, M., Keplinger, C.: Hydraulically amplified self-healing electrostatic actuators with muscle-like performance. Science **359**(6371), 61–65 (2018)
2. Biswas, S., Visell, Y.: Emerging material technologies for haptics. Adv. Mater. Technol. **4**(4), 1900042 (2019)
3. Culbertson, H., Schorr, S.B., Okamura, A.M.: Haptics: the present and future of artificial touch sensation. Annu. Rev. Control, Robot., Auton. Syst. **1**, 385–409 (2018)
4. Jiao, J., Wang, D., Zhang, Y., Cao, D., Visell, Y., Guo, X., Sun, X.: Detection and discrimination thresholds for haptic gratings on electrostatic tactile displays. IEEE Trans. Haptics **12**(1), 34–42 (2018)

5. Johansson, R.S., Flanagan, J.R.: Coding and use of tactile signals from the fingertips in object manipulation tasks. Nat. Rev. Neurosci. **10**(5), 345–359 (2009)
6. Jung, K., Kim, K.J., Choi, H.R.: A self-sensing dielectric elastomer actuator. Sens. Actuators, A **143**(2), 343–351 (2008)
7. Kellaris, N., Venkata, V.G., Smith, G.M., Mitchell, S.K., Keplinger, C.: Peano-hasel actuators: muscle-mimetic, electrohydraulic transducers that linearly contract on activation. Sci. Robot. **3**(14), eaar3276 (2018)
8. Leithinger, D., Follmer, S., Olwal, A., Ishii, H.: Physical telepresence: shape capture and display for embodied, computer-mediated remote collaboration. In: Proceedings of the 27th Annual ACM Symposium on User Interface Software and Technology, pp. 461–470. ACM (2014)
9. Liu, X., Jia, S., Li, B., Xing, Y., Chen, H., Ma, H., Sheng, J.: An electromechanical model for the estimation of breakdown voltage in stretchable dielectric elastomer. IEEE Trans. Dielectr. Electr. Insul. **24**(5), 3099–3112 (2017)
10. Mitchell, S.K., Wang, X., Acome, E., Martin, T., Ly, K., Kellaris, N., Venkata, V.G., Keplinger, C.: An easy-to-implement toolkit to create versatile and high-performance hasel actuators for untethered soft robots. Adv. Sci. 1900178 (2019)
11. Oh, C.S., Bailenson, J.N., Welch, G.F.: A systematic review of social presence: definition, antecedents, and implications. Front. Robot. AI **5**, 114 (2018). https://doi.org/10.3389/frobt
12. O'Halloran, A., O'malley, F., McHugh, P.: A review on dielectric elastomer actuators, technology, applications, and challenges. J. Appl. Phys. **104**(7), 9 (2008)
13. Oldham-Smith, K., Madden, J.M.: Electrical Safety and the Law. Wiley-Blackwell (2002)
14. Pacchierotti, C., Sinclair, S., Solazzi, M., Frisoli, A., Hayward, V., Prattichizzo, D.: Wearable haptic systems for the fingertip and the hand: taxonomy, review, and perspectives. IEEE Trans. Haptics **10**(4), 580–600 (2017)
15. Sadihov, D., Migge, B., Gassert, R., Kim, Y.: Prototype of a vr upper-limb rehabilitation system enhanced with motion-based tactile feedback. In: 2013 World Haptics Conference (WHC), pp. 449–454. IEEE (2013)
16. Shao, Y., Ma, S., Yoon, S.H., Visell, Y., Holbery, J.: Surfaceflow: Large area haptic display via compliant liquid dielectric actuators. In: 2020 IEEE Haptics Symposium (HAPTICS), pp. 815–820. IEEE (2020)
17. Simon, T.M., Smith, R.T., Thomas, B.H.: Wearable jamming mitten for virtual environment haptics. In: Proceedings of the 2014 ACM International Symposium on Wearable Computers, pp. 67–70. ACM (2014)
18. Son, B., Park, J.: Tactile sensitivity to distributed patterns in a palm. In: Proceedings of the 2018 on International Conference on Multimodal Interaction, pp. 486–491. ACM (2018)
19. Stanley, A.A., Okamura, A.M.: Controllable surface haptics via particle jamming and pneumatics. IEEE Trans. Haptics **8**(1), 20–30 (2015)
20. Suo, Z.: Theory of dielectric elastomers. Acta Mech. Solida Sin. **23**(6), 549–578 (2010)
21. Talhan, A., Jeon, S.: Pneumatic actuation in haptic-enabled medical simulators: a review. IEEE Access **6**, 3184–3200 (2017)
22. Talhan, A., Jeon, S.: Programmable prostate palpation simulator using property-changing pneumatic bladder. Comput. Biol. Med. **96**, 166–177 (2018)
23. Tanabe, K., Takei, S., Kajimoto, H.: The whole hand haptic glove using numerous linear resonant actuators. In: Proceedings of IEEE World Haptics Conference (2015)
24. Taniguchi, T., Sakurai, S., Nojima, T., Hirota, K.: Multi-point pressure sensation display using pneumatic actuators. In: International Conference on Human Haptic Sensing and Touch Enabled Computer Applications, pp. 58–67. Springer (2018)
25. Zhu, M., Do, T.N., Hawkes, E., Visell, Y.: Fluidic fabric muscle sheets for wearable and soft robotics (2019). arXiv:1903.08253 (2019)

Chapter 8
Conclusion

Abstract The longstanding goal of reproducing realistic sensations of touch, comparable to signals presented via visual or audio displays, has remained far from realization despite decades of research. One fundamental issue is that we have limited knowledge about what the hand feels during everyday interactions, considering the intrinsic coupling of movement and sensing, the complex geometry and mechanics of the hand, and the large surface area of the skin. Chapters 3, 4, and 5 apply new electronic, behavioral, and computational methods to show how simple tactile events, such as contact of a rigid object with a fingertip, give rise to dynamic spatiotemporal patterns of mechanical signals that are transmitted as mechanical waves throughout the skin of the hand. The results also show how these signals encode information about the tactile events that are involved. These ideas then informed the engineering of new technologies for haptic sensing, presented in Chap. 5, and new methods of haptic feedback, presented in Chaps. 6 and 7. This chapter summarizes the contributions of this book and discusses future research directions.

8.1 Summary

Haptics is an interdisciplinary field of research, and this book, like many in this field, contributes knowledge and methods in several different areas.

Few existing methods provide insight into the signals that are felt by the entire hand during active touch. Chapter 3 presents an investigation of these issues using a new wearable sensing method created for this work. This provided a new detailed view of touch sensing, presented for the first time in the associated publication, of how simply touching a surface elicits widespread tactile waves—spatiotemporal patterns of vibration distributed throughout the skin of the hand. This research expands current knowledge of tactile function. Prevailing descriptions of human touch sensing associate touch sensation with tactile receptors in areas of skin very near to the region of skin–object contact. This research shows how limited that prevailing view has been when applied to transient events or vibrations, and also helps explain recent findings in perception research indicating that vibrotactile signals distributed

Y. Shao, *Tactile Sensing, Information, and Feedback via Wave Propagation*,
Springer Series on Touch and Haptic Systems,
https://doi.org/10.1007/978-3-030-90839-3_8

throughout the hand can transmit information regarding explored and manipulated objects.

Chapter 4 investigated the latent dimensionality of these nominally high-dimensional tactile signals. The analysis was motivated by the structure in these signals, and neuroscience considerations, including efficient sensory encoding hypotheses. The solutions of an optimization problem formulated as convolutional sparse coding of a database of natural tactile signals yielded a dictionary of spatiotemporal primitives that provided compact descriptions (5–12 time-varying components) of information in the dataset. The representations were sufficient to accurately classify touched objects and interactions with an accuracy exceeding 95%. The primitive patterns were organized in ways reflecting the anatomy and function of the hand and the manual activities involved. Similar patterns emerged when we applied similar sparse coding criteria to spiking data from populations of simulated tactile afferents. This finding suggests that the biomechanics of the hand enables efficient perceptual processing by effecting a preneuronal compression of tactile information. The long-term implications are not yet fully understood, but future engineering influences of this work could include efficient methods for tactile sensing or feedback that leverage such low-dimensional decompositions.

In Chap. 5, we presented a new wearable instrument that extends the range of applicability of the methods presented in Chaps. 3 and 4. It consists of a 126-channel hand-wearable sensing instrument capable of capturing a wide gamut of tactile signals throughout the hand during natural manual activities. The flexible circuit geometry is adapted to the anatomy of the hand, allowing tactile data to be captured during natural manual interactions without mechanical interference. The sensing device has many potential applications, such as assisting the design and assessment of haptic interactions with commercial products, improving prosthetic and robotic sensing techniques, measuring human touch in medicine, and tactile neuroscience research. (In fact, our lab has received serious interest from current or potential collaborators in each of these areas.)

One of the great challenges in haptic engineering concerns the difficulty of mechanically stimulating the skin—a mechanical continuum with infinitely many spatial degrees of freedom. Existing practical haptic devices (ones that have been reproduced at least by multiple labs, for example) are greatly constrained in use and configuration, provide only uncontrolled and imprecise feedback, or are able to actuate the skin in a precise and controlled manner at no more than a handful of locations. A useful analogy to vision would be visual displays with only a few dozen monochrome pixels. In Chap. 6, we present a method of haptic feedback that is inspired by the wave mechanics underlying the research in the preceding chapters. We show how this method, which is based on the frequency-dependent damping of propagating waves in the skin, can be used to generate tactile stimuli with dynamically controlled spatial extent. We demonstrate that wave mechanics make it possible to achieve this using a single actuator. In psychophysical experiments and numerous demonstrations, we have found that the attributes of these mechanical signals are also reflected in perception.

Finally, Chap. 7 of this book also takes inspiration from the mechanics of the skin, a highly compliant, distributed medium, but offers a fresh approach, inspired by recent advances in emerging material technologies. In contrast to existing haptic technologies, which frequently rely on rigid actuated elements, we devised a new haptic display based on compliant, liquid dielectric actuators. This display combines electrostatic attraction with hydraulic amplification provided by a liquid dielectric encapsulated in a compliant pouch. The controlled motion of the encapsulated liquid mediates the tactile feedback that it can provide. Substantial displacements and forces are delivered across a large surface area, and the user studies demonstrate that this feedback is perceptually effective. In particular, we showed how the display can produce unique haptic experiences, such as fluid-mediated haptic effects of motion across the skin. Potential applications of the technology remain to be further explored but could range from wearable garments for social haptic communication, to patient simulators for medical training, and other HCI and virtual reality interfaces.

8.2 Future Research Directions

This book contributes to haptic science and engineering related to whole-hand tactile signals. The promising nature of the results suggests several avenues for further investigation, some of which I or my colleagues are already pursuing.

- The findings from Chaps. 3 and 4 describe a prominent role of propagating skin vibrations in touch interactions. While recently reported results in the field, and the engineering results from Chap. 6 of this book, provide some evidence for the perceptual relevance of these tactile waves, further research is needed in order to clarify this and to deduce implications for engineering design.
- Chapter 5 presented a new wearable sensing instrument informed by the results of Chaps. 3 and 4. The advantage of this wearable design is that it can accommodate diverse manual interactions, including common daily tasks involving object lift and tool use. In further developments beyond those described in Chap. 5, we developed an embedded computing platform with an integrated touch screen that interfaces with the sensor electronics described in Chap. 5 (Fig. 8.1a, showing the

a   b   c

Fig. 8.1 **a** Collecting tactile data produced during everyday hand interactions by using the portable measurement system. **b** Remote tactile feedback based on elastic wave focusing in a gelatin medium. **c** Wearable soft electrotactile display

entire system). It also allows for data storage and battery power. We plan to collect a large dataset of naturally occurring tactile signals produced during daily activities (together with other sensing methods, including video). One goal of this research is to clarify the roles played by these signals in object exploration, grasping, and manipulation. The results could inform developments in haptic science, engineering, and robotics. Similar to cases encountered in computer vision and audio, the larger size of this tactile dataset could enable refined computational methods to be applied, such as deep learning approaches.

- Chapter 6 presented a new method for rendering spatiotemporal haptic effects based on physics of cutaneous waves propagation in the hand. Although the results are promising, this study focused primarily on the fingertip, a region of the body that we often use to touch objects. The generality of the physical phenomena suggests that similar effects could be elicited at other body locations, or in multi-actuator configurations, as we hope to investigate.

- To further explore such wave-based haptic effects, my collaborators and I are investigating alternative methods for controlling waves in soft media. In recently published work [1], we show how to use time-reversal methods for focusing elastic waves via an array of remotely positioned actuators in order to produced localized tactile feedback in a soft object or interface (Fig. 8.1b). We plan to extend these computational focusing methods further in order to enable localized tactile feedback via a variety of objects and geometries, for applications in human–computer interaction, product design, virtual reality, or other areas.

- As this research shows, virtually any transient or time-varying touch contact with the skin elicits mechanical signals that become widely distributed in the skin. Thus, such processes affect a large variety of haptic feedback technologies, including some noncontact (ultrasound-based) stimulation methods [2]. One interpretation of these processes is that they decrease the feasible spatial resolution that can be achieved via localized stimulation. There is an alternative, however, which is to stimulate the skin electrically. Doing so allows small amounts of current to pass via superficial skin layers, producing palpable sensations. While these stimulation methods are challenging to apply, my colleagues and I are pursuing a new approach to engineering such a display, using a stretchable electronic circuit that provides optically addressed localized feedback. An early prototype that we have implemented is shown in Fig. 8.1c.

The generality of the phenomena investigated in this book, and the fact that this area has only recently received attention in the haptics research community, suggests that, in addition to the aforementioned research directions, the results of this book could inform a variety of developments beyond those that can be imagined today.

References

1. Reardon, G., Kastor, N., Shao, Y., Visell, Y.: Elastowave: Localized tactile feedback in a soft haptic interface via focused elastic waves. In: 2020 IEEE Haptics Symposium (HAPTICS), pp. 7–14. IEEE (2020)
2. Reardon, G., Shao, Y., Dandu, B., Frier, W., Long, B., Georgiou, O., Visell, Y.: Cutaneous wave propagation shapes tactile motion: evidence from air-coupled ultrasound. In: 2019 IEEE World Haptics Conference (WHC), pp. 628–633. IEEE (2019)

Printed in the United States
by Baker & Taylor Publisher Services